A Guide to Publishing for Academics

Inside the Publish or Perish Phenomenon

A Guide to Publishing for Academics

Inside the Publish or Perish Phenomenon

Edited by Jay Liebowitz

DiSanto Visiting Chair in Applied Business and Finance,
Harrisburg University of Science and Technology,
Pennsylvania, USA

CRC Press
Taylor & Francis Group
Boca Raton   London   New York

CRC Press is an imprint of the
Taylor & Francis Group, an **informa** business
AN AUERBACH BOOK

CRC Press
Taylor & Francis Group
6000 Broken Sound Parkway NW, Suite 300
Boca Raton, FL 33487-2742

© 2015 by Taylor & Francis Group, LLC
CRC Press is an imprint of Taylor & Francis Group, an Informa business

No claim to original U.S. Government works

Printed on acid-free paper
Version Date: 20150210

International Standard Book Number-13: 978-1-4822-5626-0 (Hardback)

Visit the Taylor & Francis Web site at
http://www.taylorandfrancis.com

and the CRC Press Web site at
http://www.crcpress.com

This book is dedicated to the next generation of academic and practitioner scholars who, through the insights and advice from the book's journal editors, will become successful in their pursuit of publications.

Contents

Preface

After spending 24 years as the founding editor-in-chief of a leading international journal and also researching in the knowledge management field, I feel it is time to "practice what I preach." I wish to share the lessons learned in journal editing with the next generation of academic and practitioner scholars who seek to overcome the "publish or perish" phenomenon. I am honored to have some of the most knowledgeable journal editors in the business and IT fields join me in contributing their knowledge in the various facets of getting published in selective, refereed international journals. It is a thrill for all of us involved in this book project to provide some helpful insights, guidance, and best/worst practices to those doctoral students, junior-level faculty, young practitioner scholars, potential reviewers, and future editors as they move forward on their research and publications journeys.

The book tries to cover the "system's view" from looking at which journal to target for one's manuscript through providing helpful tips to reviewers to contributing advice to current and future journal editors. Even though the focus is mainly in the business and IT areas, much of the guidance provided can cross other fields of endeavor through interdisciplinary research, as also encouraged in this book. In addition, we provide useful knowledge from journal editors of both traditional and online journals, as well as various tiers of journals (although favoring some of the leading journals in their respective disciplines).

We wanted to provide some interesting vignettes and do's and don'ts so that the potential author could understand what goes on behind the scenes once the manuscript arrives on the journal editor's desk (or more aptly, online submission system mailbox). We could have told some hair-raising stories about personal or professional vendettas with various colleagues regarding their research and writing activities or about plagiarism cases that consumed many months of the journal editor's time. Instead, we tried to take the "high road" and focus on key practices needed to best ensure journal publication, with some interesting cases sprinkled throughout the book.

We hope that this book serves as part of our legacy as journal editors, and provides constructive guidance for future journal authors, reviewers, and editors. It certainly was a catharsis to assemble this book, and we feel that you should derive value-added benefits from applying the knowledge in this book.

We wish to thank our families, students, colleagues, publishers, readers, authors, reviewers, and editors over the many years that we have served as journal editors in helping us shape our way of thinking. Doing "journal patrol" (as I call it) is a daily time-consuming activity, but in the end, we feel we have an impact on helping one's career (we hope, a positive one). Getting an e-mail from a reader who said that the journal has greatly shaped her research agenda or the newly minted assistant professor who eventually has his dissertation research published makes serving as a journal editor a rewarding experience.

We hope you enjoy the book and will pass these "gems of knowledge" to your students, colleagues, and others.

Jay Liebowitz, DSc
Harrisburg, PA

List of Contributors

Donald E. Brown
Editor-in-Chief
IEEE Transactions on Systems,
Man, and Cybernetics,
Part A: Systems and Humans

John S. Edwards
Founding Editor-in-Chief
Knowledge Management Research
and Practice Journal

Steven R. Gordon
Former Editor-in-Chief
Journal of Information Technology
Cases and Applications Research

Jeremy Hall
Editor-in-Chief
Journal of Engineering and
Technology Management

Murray E. Jennex
Editor-in-Chief
International Journal of
Knowledge Management

Alex Koohang
Editor-in-Chief
Journal of Computer Information
Systems

Jay Liebowitz
Founding Editor-in-Chief
Expert Systems with Applications:
An International Journal
and
Editor-in-Chief
Procedia–CS

Dennis E. Logue
Co-Editor
Financial Management
and
Editor
Journal of Corporate Finance

James R. Marsden
Editor-in-Chief
Decision Support Systems Journal

Sibel McGee
Managing Editor
Journal of Homeland Security and
  Emergency Management

Daniel E. O'Leary
Editor-in-Chief
International Journal of Intelligent
  Systems in Accounting, Finance
  and Management
and
Former Editor-in-Chief
IEEE Intelligent Systems

Joanna O. Paliszkiewicz
Deputy Editor-in-Chief
Management and Production
  Engineering Review

Irmak Renda-Tanali
Editor-in-Chief
Journal of Homeland Security and
  Emergency Management

Suprateek Sarker
Editor-in-Chief
Journal of the AIS

Anthony K.P. Wensley
Editor-in-Chief
Journal of Knowledge and Process
  Management

Duane Windsor
Editor-in-Chief
Business & Society

Arch G. Woodside
Editor-in-Chief
Journal of Business Research

1

Don't Do as I Do; Do as I Say

JOHN S. EDWARDS

Contents

Introduction

This chapter reflects on two journeys that have taken place in parallel: a 40-year academic career including a substantial amount of time spent mentoring other academics, and a 15-year adventure in launching and running a journal in a newly emerging field (knowledge management, KM) at the intersection of several disciplines. My intention is to draw some lessons (12 in all) about publishing research articles, including both writing and reviewing, that help researchers to publish in a way that furthers their career, especially those still near the beginning of it.

The title of the chapter reflects both the changing nature of academic careers and assessment of academic performance over time, and the fact that the right or best publication strategy for an early career academic is not necessarily the same as that for a full professor with 25 years' experience.

Let's start with my own background, so you know where I am coming from, as the phrase goes. My career has followed a fairly standard academic trajectory, although I did have two or three years' experience of "real" work before and during my PhD studies. In terms of subject, I seem to have become gradually "softer" over the years, starting with a mathematics degree, then a management science PhD, then moving from mathematical model-based management science/operational research into expert systems, knowledge-based systems, decision support systems, business processes, and finally (so far?) knowledge management. I have published more than 70 refereed journal articles, several in top journals, and reviewed for around 20 different academic journals. I have also edited several special issues of journals, and helped found the journal *Knowledge Management Research & Practice (KMRP)*, spending 10 years as its editor from 2003 to 2012.

As well as research (and teaching, which is not our concern here), other people seem to feel I have expertise in academic management, so I have held several such positions, culminating in being executive dean of a business school. This means I can bring to this chapter the perspectives of an experienced published researcher, a journal editor and reviewer, and someone who for many years has advised and indeed has had to judge the performance of junior, and not-so-junior staff. My own experience has been in the United Kingdom, but most of it in a school where nearly half the academics come from other countries, so with that and my network of international contacts, I trust that most of my contribution applies irrespective of country.

The crucial warning for any older academic—like me—trying to advise younger colleagues is to try to remember what it felt like to be a new academic, whether new through age, as in my case, or after a career change, and also to weave that with the realities of now. I *can* still recall being a young academic, although I have mentored many colleagues over the last 20 to 25 years. My advice seems to have worked for the vast majority of them, so here goes.

Plus ça Change ... ? (The More Things Change ... ?)

Times have changed since my academic career began in the 1970s. There is more emphasis on publications now, especially publications in top journals. But it's not as straightforward as that. When I started my academic career, on the path to tenure, yes, publications in refereed journals were essential for promotion/tenure, but what we might call the "conveyor belt" expectation regarding publications was not yet widespread. So in any given year, refereed journal articles were only a "nice to have." Other forms of dissemination of research work, such as conference papers and presentations, also had more of a role in judgments of an academic's performance. Even conference attendance was seen as useful, for what nowadays we call networking.

There have been both internal and external drivers for the changes that have happened since the 1970s. There has been a greater professionalization of the management of universities, generally seen as a good thing by external commentators, but in some cases leading to a divergence of views internally. A business school that did not advocate a professional approach to its own management would be a contradiction in terms, but that needs to be achieved without losing the collegiality for which universities have been renowned (Miller, 1994). Externally, there is also now much more formal assessment of all aspects of a university's performance, including research. The UK government, for example, first began formal assessment of the quality of university departments' research in 1986. Other rankings and league tables have also proliferated, and even teaching rankings usually include some evaluation of research as well; for example, the influential *Financial Times* ranking of MBA programs, first published only as recently as the end of the 1990s, judges research quality on the basis of the number of publications in an "elite list" of journals.

Most relevant to this chapter are journal ranking lists, telling us all which journals are A, A*, 4* (and lower categories!). If "official" rankings for business journals existed in the late 1970s, I can only say that I was not aware of them, nor were the professors who ran my department. The ingredients were there, though: impact factors were first published for journals in the science category of the *Journal Citation Reports* in 1975, and the journal *Scientometrics*, which specializes in the field, was founded in 1978. Whether the same ranking approach is appropriate in different subjects is still a matter of much debate. It is noteworthy that Eugene Garfield, coinventor of the method of calculating impact factors and founder of what is now Thomson ISI, the publishers of the *Journal Citation Reports,* always seems to have used examples from science rather than social science (let alone specifically business) in his presentations.

Up to the 1980s, the few rankings of business journals that were published were based on opinion surveys, generally of the perceptions of departmental chairs or heads. For example, see Coe and Weinstock (1984, data collected in 1982) for a ranking of management journals. This actually replicated survey work they had published in 1969 (using 1968 data), although ironically that did not appear in a refereed journal, but in the *AACSB Bulletin.*

Nowadays, there are all kinds of journal ranking lists produced by institutions, subject areas/associations, and even with official status at the national level (e.g., in Australia and Norway). In business, the Association of Business Schools' ranking list has been very influential over the past decade, especially in the United Kingdom. That ranking has now been subsumed into a larger project, boldly entitled, *The International Guide to Academic Journal Quality,* even though around three-quarters of the scientific committee members are based in the United Kingdom (see http://www.bizschooljournals.com/). I await the promised 2015 first edition of this *Guide* with interest and some trepidation!

So, what do all these lists and rankings mean for you as an early career academic? Above all, despite the implications of all these lists, attitudes to publication do vary between institutions, even between departments in the same institution, and not in what might be thought of as the obvious ways. For example, many institutions and deans are indeed obsessed with journal ranking lists, article citation counts, and impact factors. But it is the middle-ranking or aspiring institutions that tend to use these in the most mechanistic way for promotion, appraisal, or tenure decisions. The higher-ranking institutions assess the individual papers themselves (or invite a further peer review). The lower institutions don't care! That leads to:

Lesson 1

Find out if your department uses a journal ranking list in promotion/ tenure/appraisal decisions. And if it does, make sure you have a copy of it. But use the information on it with care (see later).

Should I Publish at All? It may not have occurred to you to question whether you should even try to publish a particular piece of research. Many academics still believe, quite reasonably in my view, that any of their work which may be useful to someone else ought to be published somewhere. This can lead to a curriculum vitae (CV) with a large number of publications in a wide variety of journals. What's wrong with that, you may ask? Yet some reviewers and promotion commit- tees criticize a candidate with this sort of CV even if there is a core of top journal publications in there. Their argument is that all the extra publications dilute the focus on the academic's main area of expertise. I know what you think … why would you ever want to *not* publish your research? But some academics would retort that you shouldn't even *do* research that would not be publishable in a highly rated jour- nal, because of the opportunity cost.

There's an implication in that latter view that publication in top journals is the only purpose for academic research. An alternative view is that the impact of research outside academia is also a worthy purpose. At the moment, the latter view is gaining ground, especially in the United Kingdom, where in 2014 for the first time research impact has been built into the government assessments of departmen- tal research quality (results eagerly awaited at the time of writing). In the end, it depends on what your department or institution values, and perhaps to some extent your own preference.

In the rest of the chapter, I assume that you choose the journal and then write the paper. Perhaps I should say "the paper proper," as it's likely to exist in some form before you contemplate submitting to a journal, especially if you have already produced conference papers on the topic, which is a good idea.

Choosing a Target Journal

So, let's assume from now on that you want to publish your research, either to help gain promotion/tenure/a good performance appraisal or simply to tell other people what you have done. It's fair to assume also that you will always want to publish in the best journal possible. However, the definition of "best" will vary according to your own purpose in publishing. This is a key "Don't do as I do" point: generally (although not always) the top journals, in the sense that the journal ranking lists purport to measure, have a longer lead time from submission to publication than lesser journals, because of their more rigorous review processes and perhaps also the number of papers submitted to them. I have seen very junior academics who have sent their first papers (typically the ones from their PhD thesis) to very top journals (advised by a full professor for whom that would be a sensible strategy), with review processes lasting more than a year and then two years from acceptance to publication, only to be castigated for not getting any papers accepted in their first year in post. The trend toward accepted papers first appearing online, well in advance of the printed journal issue, has reduced this problem somewhat, but it still exists.

Even the various rankings disagree considerably on how many top journals there are worldwide. The *Financial Times* list includes 45 journals; the Australian government rankings in 2010 listed 24 journals in the top category (A*) for business and management; the equivalent Association of Business Schools list for 2010 rated 93 journals in its top (4*) category. Whatever the number may be, we can't all publish in the very top journals, no matter how good our research is. With more than 12,000 business schools worldwide and that number increasing by the month, you can see that there simply isn't room for that many articles.

Taken to its logical conclusion, the mechanistic approach would of course mean that no new journal could ever be launched successfully.

The best advice I can give for a junior academic in targeting a journal is to talk to your more senior colleagues and mentor(s). Academic disciplines are all about conversations. That is what the journals and conferences are: a way of carrying on conversations between people interested in or working in the field. Indeed, it has been argued that all management is about dealing with conversations (Winograd and Flores, 1987). So here is:

Lesson 2

Which conversation are you seeking to join by publishing your research? (What is your research about, at the most fundamental level?) And what are you intending to contribute to the conversation? (Why would people want to listen to you?)

Once you have decided that, you need to go further into detail. From Lesson 1, you'll have found out which ranking list(s) matter, if any, in your institution. Talk to the senior academics and colleagues who advise you, and find out how important the lists really are to your future. If decisions in your institution are not list-based, find out what your advisers think are the best journals possible for your work. And that plural—journals—is important. Journals regarded as being of international standing, not just the top ones, but down to the B or 2* journals, generally reject far more papers than they accept, so you need a second and possibly even third choice as a fallback. That leads to:

Lesson 3

Choose the best journal possible for your paper, and also decide what your fallback journal(s) will be if your paper is rejected by the first choice. This goes in parallel with:

Lesson 4

Know what game you are playing.

The better the journal's prestige or reputation, the more important this is. What are the journal's expectations in terms of the way that a paper is constructed? This can apply to length, topic, academic content, datasets, writing style, and even methodological approach.

The point about methodology is very important for early career academics, especially if your discipline is one that suffers from a "quantitative or qualitative" methodological split, as in (for example) information systems, operational research, and psychology. If it does, there may well be journals that literally will not even review a qualitative study (the reverse is less common). And if you take a mixed methods approach in such a discipline, one using both quantitative and qualitative methods in the same study, you need to choose your target journals even more carefully.

In extreme cases, the expectations even apply to the whole process of developing a paper for the journal. And sadly, this game is not always on a level playing field. One top journal which had better remain nameless has a paper development process that effectively requires a version of the paper to be presented at a workshop, and these workshops are nearly always held in the United States, at a time in the year that suits North American attendees well but European academics very badly.

If you are still uncertain about something:

Lesson 5

If in doubt, ask the editor.

As a junior academic, you might think that you shouldn't waste the time of busy senior people with what might be trivial inquiries. Actually the opposite is true, at least of journals. If your paper does fit, most editors will welcome the initial contact as a chance for clarification on both sides. If it does not fit, then you will have saved everyone's time, including the editor's and reviewers' as well as your own.

"Fast Track" Routes   Two topics come under this heading: conferences that advertise routes to publication in one or more journals, and journal special issues, that cover a specific topic within that journal's scope and have one or more guest editors. A conference "fast track" route is fine, but unless speed of publication is your main criterion, don't let the existence of such a route seduce you into submitting to a journal that is not one of the "best" for your research. As for special issues, if the topic of a special issue of a journal really suits the research that you want to publish, then go for it; special issue articles often have a shorter lead time than regular papers. But don't attempt to skew your work so that it appears to fit the special issue, and again, don't submit to a journal that you would otherwise not consider.

Right, you've chosen your target journal(s). Mention of writing style earlier links us neatly into the next section.

Writing the Paper

Before you even start:

Lesson 6

Read the instructions to authors for the journal, and more important, follow them!

These instructions can cover any or all of the organization of the paper, the methodological approach taken, the format and style of the paper, especially the references, and the mechanisms for submission (increasingly, but not always, online).

In terms of format and style, one thing that has certainly changed over 40 years is the extent to which authors now are supposed to worry about fonts, type size, line spacing, margins, and so on. When I started it was typically, "Submit three copies of your manuscript, typed double-spaced with wide margins." But believe me, the benefits of word-processing software far outweigh any inconvenience this causes! I won't name the journal whose instructions still assume you are submitting your word-processed manuscript by mailing them a floppy disk.

As for references:

Lesson 7

Use bibliographic software to manage and format your references. There may be an initial investment of effort in doing this, but it is well worth it.

Returning to Lesson 2, let's extend Lesson 6 a bit:

Lesson 8

Read the relevant papers in your target journals.

Then you will know what the kind of paper you are writing generally looks like. As a researcher, you should know this already, but I have sometimes as an editor received papers that seem to be written by people who have never read any academic journal, let alone the one to which they were submitting.

This also relates to a much-misunderstood point about citing papers from the journal to which you are submitting. Yes, there are a few journals (but not top ones) that have a mechanistic insistence on doing this in a misguided attempt to bump up their impact factors. However, remembering again that journal articles are part of the discipline's conversation, we can see that this is a meaningful issue: why publish it here, in this journal? Why do you want to join *this* conversation if nothing that others have said in it is worth mentioning? Good practice for both journals and authors, in my opinion, is to identify a core group of similar journals; then it is certainly reasonable to insist on references from that group of journals. For *KMRP*, this might be (say) the *Journal of Knowledge Management*, the *International Journal of Knowledge Management*, *VINE*, and maybe *Knowledge and Process Management*. But again, and let's be frank here, if all your references are to one particular journal, and you submit it to a different one, the editor may have good reason to suspect that the paper you have submitted has already been rejected by that particular journal!

As for which papers you should cite, generally my advice would be to decide on the basis of academic content rather than what you think the "best" ones are. There is one exception to this: for definitions and any other essential foundations of your paper, you need to cite a paper in a journal of international standing (at least B/2*) or a book by a well-known and published researcher. Not (say) a conference paper, a University of Poppleton working paper, or a paper from the *Erewhon Journal of Knowledge Management*. Beyond that, it's a matter of judgment. Citation counts are a dubious measure of quality, even of an individual paper. I once edited a special issue of a 3* journal. The most-cited paper in the issue to this day was in my view the weakest paper of the lot; it got in by the skin of its teeth. But it had an engaging *title*, making good use of key buzzwords. So, that's another lesson:

Lesson 9

Make sure your paper has a good title, which means that it is both a clear title and an accurate one.

In these days of Internet search engines that would enable you to identify the authors in seconds, the only examples I dare give here are my own. "Using Grounded Theory for Theory Building in Operations Management Research: A Study on Inter-Firm Relationship Governance" (Binder and Edwards, 2010) can hardly be bettered, although it is a bit long. "Knowledge Management Systems and Business Processes" (Edwards, 2005) is clear and accurate, but maybe could be a bit more precise, for example, "The Relationship between…". But "Health Visitors Revisited" (Edwards et al., 1983)? It doesn't really give any clue as to what our research findings might be, does it? I was a lot younger then, of course.

And finally, a lesson I wish I didn't have to include:

Lesson 10

Don't copy other people's work. You *will* be found out.

Any decent journal these days routinely uses the same plagiarism detection software that is used to check student assignments. And yet you might find it hard to believe what some people will do. In my time as editor of *KMRP*, I twice received papers that were completely copied from papers previously published by other authors in another KM journal, one absolutely unchanged, the other with two pages of newer literature tacked on to the end of the literature review. Needless to say we didn't take them any further. The authors have been barred from further submission and not just to *KMRP*, but to all Operational Research Society journals and other reputable KM journals. In one case, I also wrote to the author's boss at his institution pointing out what the junior member of staff had done (although I never received a reply). I couldn't do this in the other case, however, because the culprit *was* the boss of his institution, a stand-alone business school.… If I hadn't seen it happen, I don't think I would have believed that story myself!

If that's your approach, you won't be reading this book, but readers, please do make sure that when you are using definitions or results from other authors' work, you put them in quotes and cite them fully. Otherwise you might set off the journal's plagiarism detection software warning flags.

Lastly, before you submit, make sure someone else has read the paper, both for content and for proofreading. If your first language is not the one the journal is published in, then find out if the journal offers language support for nonnative authors. If not, you'd be well advised to get your paper read by a native speaker or a teacher of the relevant language.

What Happens Next? Reviewing

My view, stemming from seeing journal articles as part of a conversation that moves a discipline forward, is that reviewing should be first and foremost a constructive process, not one that's based on "catching the authors out." An author who submits to a journal should learn from having his or her paper reviewed, even if the paper is rejected. This has implications for the tone of reviewer comments as well as their content. My take is that reviewers should not write something in a review that they would not be willing to say to the author's face. OK, there are some people who would be willing to say things face to face that most of us wouldn't; this happens occasionally in the question and answer session at the end of a conference presentation, and you can see the other people in the room cringing or hear the collective sharp intake of breath when one of these "aggressive academics" goes for the jugular of the poor presenter.

Some people feel all papers should have the same structure. But does this hinder novelty? It may well be easier in a laboratory experiment-based science discipline than in business. Unless a structure is specified by the journal, then this is all part of the academic conversation.

There is certainly variability in the reviewing process. I have been editor for papers that have been accepted first time with minimal changes, and alternatively some that have had three full rounds of review before acceptance. I've seen the same as a journal article author, and I'm not talking about different practices between different journals here, but what we might call "within-journal" variation.

The crucial point to note is:

Lesson 11

Remember that the review process includes an element of discussion.

If the initial decision is "revise and resubmit," do not be afraid to disagree with what the editor/reviewers have said, as long as you can (solidly) substantiate your position. They are human, after all! And when you submit the revised version, explain exactly what changes you have made (or not made) in response to the editor and reviewer comments.

However, in parallel with this there is also:

Lesson 12

Although Lesson 11 says that it's OK to engage in debate with the editor or reviewers about the content of your paper if you are sure of your ground, don't argue with the editor about the processes of their journal. They will have used these processes about a hundred times more than even the most published of authors, never mind an early career academic.

I once had an author object to making certain changes on the grounds that "the editorial team could do this"—something I have never known to happen in my own career—on the grounds, as they said, that "from my experience, most journals would not ask the author to do this." And this from an author whose journal publications, including this one, could be counted on the fingers of one hand.

The most common reason for rejection of "half decent" papers in my experience is failing to link the theoretical and empirical parts of the paper properly. So there might be a reasonable literature review about some aspect of knowledge management, and reasonable collection and analysis of some data relevant to knowledge management, but the paper fails to tell me why analyzing that specific data, rather than some other similar or related data, adds to the conversation. Often this is because the data collection came before the full literature review was done, on the basis of a hastily constructed survey or set of interview topics, and that's a fault that will almost certainly be impossible to rectify.

Final Remarks

When your paper has finally been accepted, it's not quite all over. The proofs will come to you for checking before publication. Make this a top priority and do it carefully: only you know what you meant to say, and it's your responsibility to make sure that what goes out into the world is correct. Once you've done that, then you will be able to look at your publication and think, "The effort was worth it."

And don't let rejections get you down. As I said earlier, far more papers are rejected than accepted, and with double-blind review even experienced researchers still find their work rejected (I know I do!).

Good luck with your publishing career!

References

Binder, M. and Edwards, J.S. (2010). Using grounded theory for theory building in operations management research: A study on inter-firm relationship governance. *International Journal of Operations & Production Management*, 30(3): 232–259. doi:10.1108/01443571011024610

Coe, R. and Weinstock, I. (1984). Evaluating the management journals: A second look. *Academy of Management Journal*, 2(3): 660–666. doi:10.2307/256053

Edwards, J., Luck, M., and Medlam, S. (1983). Health visitors revisited. *European Journal of Operational Research*, 14(3): 305–317.

Edwards, J.S. (2005). Knowledge management systems and business processes. *International Journal of Knowledge and Systems Sciences*. JAIST Press, 2(1): 10–18.

Miller, H. (1994). *The Management of Change in Universities*. Maidenhead, UK: Open University Press.

Winograd, T. and Flores, F. (1987). *Understanding Computers and Cognition: A New Foundation for Design*. New York: Addison-Wesley.

2

GRABBING READERS

How to Focus Your Paper's Title and Contents
on Its Major Theoretical Contribution Rather
than the Local Context of the Study

ARCH G. WOODSIDE

Contents

Introduction

Which of the following two titles did the author(s) of the journal article or book actually use: (A) *Client–Hairdresser Conversations in the Beauty Salon Industry in Metropolitan Regions of Turkey* or (B) *How Marketplace Performances Produce Interdependent Status Games and Contested Forms of Symbolic Capital* (Üstüner and Thompson, 2012)? Here are two additional titles to select one from: (A) *Meetings and Pastimes of Young Male Friends in the North End of Boston* or (B) *Street Corner Society* (Whyte, 1943). This chapter offers a primer on writing titles and relating the scholarly content of your academic paper to contribute original theory to a discipline and get your paper accepted for publication. The chapter offers tenets—sets of helpful rules for writing titles and paper content—for increasing readers' interest and editors' acceptances of authors' paper submissions for publication in SSCI journals.

This chapter focuses on 10 useful rules of thumb to consider when writing a title for a paper that you plan to submit for publication consideration to an editor of a scholarly journal in the behavioral sciences including more than 16 subcategories of journals in the general field of business, economics, and management. Combining these 10 rules provides an algorithm (a checklist; see Gawande [2010] on the value of checklists) designed to be useful for increasing clarity and the attention, understanding, and acceptance that your paper receives by the editor and, one hopes, reviewers.

Rules

Rule 1: Ask Your Coach If the Title Is the Right One for Your Paper

According to Gawande (2011), each of us needs a great coach who offers insights and reviews our actions *in situ*. "*In situ*" refers to the specific context, the specific paper that you are about to submit to a journal editor. A great coach is a coach who is helpful and has had several successes herself or himself in writing papers appearing in major journals in your field. If you are an early career scholar, consider seeking a qualified coach. A useful metric for a qualified coach is a coach having an i10-index greater than 10; that is, the coach has written 11 or more publications each having 10 or more citations. A senior colleague at your college or university who you are comfortable asking for advice often makes a great coach.

Do not be so overconfident in the quality of the title of your paper that you submit it without asking your coach. Is the title the right one for your paper? Of course, also ask your coach to read the paper and provide comments. Remember to thank your coach for her or his suggestions in reading a draft of your paper; do so on the bottom of the title page of your paper. Not only does applying this rule (almost) always improve the quality of the title of your paper and its contents, usually editors consider papers having such acknowledgments to be higher in quality than papers giving no indication of such early reviews. The first tenet is to recognize that you will improve your thinking and writing via coaching as master-writer Gawande (2011) informs us. The title to the present chapter offers a hint for answering the two title-quiz questions correctly: focus your main title on the paper's major theoretical contribution.

Rule 2: Read the Articles/Titles in Your Field
Having the Highest Citation Impact

The implicit tenet for this second rule is that high-impact papers more often than not have great titles. Adopting this perspective for a moment, go to Harzing.com's "publish or perish" software and click on "journal impact" and type in one of the high-citation impact journals in your discipline. Read the titles of the top 20 articles by citation impact appearing in the journal.

Rule 2 raises the question of what journals to search to identify relevant journals. Early career scholars experience a number of difficulties in writing papers for publication in acceptable scientific journals. "Acceptable" refers to journals in prestige rankings of academic journals; Harzing (2014) provides journal rankings for 22 ranking studies. Among the listings in Harzing (2014), for example, is the WIE 2008, the *WU Wien Journal* Rating May 2008 List developed by the Wirtschaftsuniversität Wien (Vienna University of Economics and Business Administration); the 2008 list of journals only contains journals with an A+ and A ranking. The *Financial Times* Top 45 Journals Used in Business School Research Rankings is another useful ranking of elite business-related journals (FT, 2014). The journal

Table 2.1   Google.com/scholar Journal Listing by Article Citation Impact

RANK	JOURNAL	H5-INDEX[a]	H5-MEDIAN
1	*Journal of Marketing*	65	111
2	*Journal of Business Research*	62	84
3	*Journal of Consumer Research*	51	68
4	*Journal of Marketing Research*	50	76
5	*Industrial Marketing Management*	47	69
6	*Journal of the Academy of Marketing Science*	45	65
7	*Journal of Product Innovation Management*	43	61
8	*Marketing Science*	42	65
9	*European Journal of Marketing*	36	50
10	*Journal of Retailing*	34	59

Source: http://scholar.google.com.au/citations?view_op=top_venues&hl=en&vq=bus_marketing (accessed September 2, 2014).

[a] The h5 index is the h-index for articles published in the last five complete years. It is the largest number h such that h articles published in 2009–2013 have at least h citations each; thus the *Journal of Marketing* has published 65 articles with each of these articles having 65 or more citations. The h5-median is the h5-median for a publication that is the median number of citations for the articles that make up its h5-index.

rankings by the Australian Business Deans' Council (ABDC, 2013) is a third list; this list ranks journals into four levels of prestige: A* (A star), A, B, and C; take note that faculty staff members' publications appearing in C-ranked journals receive zero credit or minus credit for such articles at some of the 40 universities in Australia. The Google.com/scholar journal listing for articles' impacts is the fourth and final listing receiving mention. Google provides journal listings for articles' impacts for 16 subcategories of "Business, Economics & Management." For example, Table 2.1 contains the top 10 Google journals by impact for the marketing subcategory. Consider taking time to download listings of the titles of the top-cited articles in a few of these journals.

Using a search of *The Journal of Marketing* (*JM*) at Harzing.com, Table 2.2 is a list of the top 20 articles by number of citations appearing in the *JM*. What do you notice about the titles in this list? The following words appear a few times each: service, model, relationship, conceptualizing, consumer, and theory. Although the second rule is *not* to suggest using words appearing in the top-cited articles in prestige journals, do consider focusing your title on the theoretical contribution that your paper is making and not on the specific context of the study that your paper reports. Ask yourself the following two

Table 2.2 Top Twenty Articles by Citation Impact Appearing in *The Journal of Marketing*

CITES	PER YEAR	RANK	AUTHORS	TITLE
☑ h 14825	511.21	2	A Parasuraman, VA Zeithaml, LL Berry	A conceptual model of service quality and its implications for future research
☑ h 14083	704.15	1	RM Morgan, SD Hunt	The commitment–trust theory of relationship marketing
☑ h 9378	360.69	3	VA Zeithaml	Consumer perceptions of price, quality, and value: a means-end model and synthesis of evidence
☑ h 8280	306.67	4	FR Dwyer, PH Schurr, S Oh	Developing buyer-seller relationships
☑ h 8134	387.33	6	KL Keller	Conceptualizing, measuring, and managing customer-based brand equity
☑ h 8083	367.41	5	JJ Cronin Jr, SA Taylor	Measuring service quality: a reexamination and extension
☑ h 6900	287.50	7	JC Narver, SF Slater	The effect of a market orientation on business profitability
☑ h 6765	281.88 *	8	AK Kohli, BJ Jaworski	Market orientation: the construct, research propositions, and managerial implications
☑ h 6559	364.39	9	VA Zeithaml, LL Berry, A Parasuraman	The behavioral consequences of service quality
☑ h 5804	276.38	10	BJ Jaworski, AK Kohli	Marketing orientation: antecedents and consequences
☑ h 5397	224.88	11	JC Anderson, JA Narus	A model of distributor firm and manufactuer firm working partnerships
☑ h 5343	267.15	12	S Ganesan	Determinants of long-term orientation in buyer-seller relationships
☑ h 5165	303.82	13	PM Doney, JP Cannon	An examination of the nature of trust in buyer-seller relationships
☑ h 5136	342.40	15	RL Oliver	Whence consumer loyalty?
☑ h 5050	252.50	14	GS Day	The capabilities of market-driven organizations
☑ h 4803	218.32	16	C Fornell	A national customer satisfaction barometer: the Swedish experience
☑ h 4555	253.06	17	DL Hoffman, TP Novak	Marketing in hypermedia computer-mediated environments: conceptual foundations
☑ h 4554	227.70	18	EW Anderson, C Fornel, DR Lehmann	Customer satisfaction, market share, and profitability: findings from Sweden
☑ h 4286	142.87	538	C Grönroos	A service quality model and its marketing implications
☑ h 4248	177.00	19	MJ Bitner	Evaluating service encounters: the effects of physical surroundings and employee responses
☑ h 3938	164.08	20	MG Bitner, BH Booms, MS Tetreault	The service encounter: diagnosing favorable and unfavorable incidents

Source: Harzing's publish or perish (accessed September 2, 2014).

questions. What is the contribution to theory that your paper is making? Does your title indicate that contribution?

Notice that only one paper refers to a country by name (Sweden) among the 20 titles appearing in Table 2.2. Unless your paper is to test a theory across several countries avoid a country, state, city, or place in your title. Focus the title on identifying the theory and the theoretical contribution that the paper delivers. If you download the most cited articles in each of the top 10 journals by citation impact appearing in Table 2.1, you will find "model" and "theory" or "theoretical" appearing more often than other words in the first article of each list. Be sure to answer in the title and especially the paper the question, "What theoretical contribution is your paper making to the discipline?" Your title should respond explicitly to the implicit conclusion that many editors and reviewers are prone to reach when reading a submission: the contents of this paper are relevant only to the context in the author's study. Show how the study in your paper is generalizable to theory as well as generalizable to contexts beyond the one that you report.

Rule 3: Use Two Titles—The First Five Words or Less

The main title might state the topic and the subtitle an implied or explicit proposition. Gawande's (2010) checklist title illustrates such a title: "The Checklist. If Something So Simple Can Transform Intensive Care, What Else Can It Do?"

The main title might state an action and the subtitle one or more paradoxes that relate to the main title. If an editor insists, the authors could delete the subtitle due to length restrictions.

Rule 4: Avoid "Replication" and "Extension" in Your Title

Replication studies are highly valuable, but gaining acceptance of such studies is difficult, and the difficulty is increasing, according to Evanschitzky et al. (2007). Reviewers and more than a few editors falsely believe that a second or third study with findings confirming those in an earlier study provide no new information (Evanschitzky and Armstrong, 2013). Independent confirmation of findings is the most powerful method available for establishing accuracy. However, a consistent negative bias exists in the review processes of many academic journals toward replications.

Thus, a paradox exists! Replications are highly valuable reading that many reviewers and editors perceive to have little value. Resolving the paradox and achieving success in getting replications accepted for publication include taking several steps. The steps include start framing the title of your paper with "updating" or "supporting" or "extending" or "refining" or "additional testing" rather than "replicating." A synonym for "replicating" is "duplicating," and "duplicating" is close to "copying" which is not far away implicitly from "plagiarizing." The suggestion is that an unintended consequence of using "replication" is "nothing new" at best and at worst stealing.

A second step is to read the literature supporting the usefulness and how to do replications. Uncles and Kwok (2013) is solid fundamental reading on the value of replications and how to do them well. When offering a replication study, be sure to cite and discuss three to six articles that describe the high value of replication research and their near total lack of appearance in journals (a third step). A fourth step is to build in replications into your own original studies. Uncles and Kwok (2013, p.1403) advocate the "the replication process should be integral to the design of the original research project—not something dismissed as 'a limitation of the study' for others to resolve at a future date."

Rule 5: Avoid Openings with Low Information Content:
Words to Avoid Include "Critical," "An Empirical Analysis of,"
"The Major Determinants of," and "The Effects of"

Great titles get to the issue very quickly. Here is a great title, "Focusing on the Forgone: How Value Can Appear So Different to Buyers and Sellers." Note that this title starts with a paradox that violates the old saw, "Out of sight, out of mind." The second half of the title offers a promise to answer the riddle of how value can appear to be so different between buyers and sellers. Ziv Carmon and Dan Ariely (2000) are brilliant at writing great titles as well as doing and writing up great studies.

Rule 6: Describe a Paradox

A paradox is an inconsistency, an irony, a contradiction, or an illogicality. Readers continue reading when an author presents a paradox.

"Rational irrationality" is another paradox. How can irrationality be rational? Go to Google.com/scholar; enter "rational irrationality" and over 962 answers or explanations will appear. Readers love to read quandaries and ruminate on solutions to quandaries.

Rule 7: After Achieving over 5,000 Citations,
Write Your "The General Theory" Paper

A book by John Maynard Keynes (1936), *The General Theory of Employment, Interest and Money* has nearly 24,000 citations, a book that has had a huge impact in economics and public policy! In addition, Google.com/scholar includes 433,000 results for "the general theory of." Offering a general theory of anything commands a lot of attention. Usually your general theory should offer different theories and explain specific context where each context applies within your general theory.

Rule 8: Use an Intriguing Turn that Generates a Vision: "Gone with the Wind,"
"The Presentation of Self in Everyday Life," and "Street Corner Society"

Measured by citation impact, *The Tourist Gaze* (Urry and Larsen, 2011) is the most intriguing turn of phrase in the business/management subdiscipline of hospitality and tourism. *The Presentation of Self in Everyday Life* (Goffman, 1959) is another exceptional, delightful, and high-impact turn of phrase. *Street Corner Society: The Social Structure of an Italian Slum* was William Foote Whyte's (1943) first book (and his PhD dissertation at the University of Chicago) and is a third great turn of phrase. Note the effective use of two titles by Whyte. His book is a participant observation study about the life in the North End of Boston occurring during the Great Depression. The book that is a compelling read often follows a great title.

Rule 9: Write Three Possible Titles and Show Them to Three
Colleagues Separately and Ask, "Which One Can I
Get You to Read and Comment on for Me?"

Who applies this rule? Use a "behavioroid" step in implementing this rule. Instruct your participants that you have found three books at Amazon.com, each with one of these three titles. "I wish to buy one or more books for you. The price is about the same for each book. I will order one book for you today. Which one should I buy for you? (I am sorry that I can only tell you the title of each book.) Which one do you want as a birthday gift? Please pick one and please think aloud as you go about selecting your present." Write distinctly different titles in preparing this game (test). Frequently your least favorite will be most chosen among 10 people who you ask to participate in the study.

Rule 10: List a Sequence of Acts for Your Title

Eat, Pray, Love is a great title. An unexpected runaway hit for the author (Gilbert, 2006) and then a successful movie. I am expecting the same for "Embrace·Perform·Model: Complexity Theory, Contrarian Case Analysis, and Multiple Realities" (Woodside, 2014). Note the subtitle, just to be clear about the main topics of the article.

Conclusion

Titles vary in quality and impacts just as articles do. Take the time and make the effort to try several or all of the rules appearing in this article. Doing so is likely to improve the quality, acceptability, and the impact of your article.

Acknowledgment

My thanks go to Samir Gupta, Monash University, for service as my coach for crafting the title of this chapter. I thank Carol M. Megehee, my wordsmith coach, for comments and revision suggestions on an early version of the chapter.

References

ABDC (2013). http://www.abdc.edu.au/pages/abdc-journal-quality-list-2013.html (downloaded, November 17, 2014).

Carmon, Z. and Ariely, D. (2000). Focusing on the forgone: How value can appear so different to buyers and sellers. *Journal of Consumer Research*, 27(3): 360–370.

Evanschitzky, H. and Armstrong, J.S. (2013). Research with in-built replications: Comment and further suggestions for replication research. *Journal of Business Research,* 66(9): 1406–1408.

Evanschitzky, H., Baumgarth, C., Hubbard, R., and Armstrong, J. (2007). Replication research's disturbing trend. *Journal of Business Research*, 60: 411–415.

FT (2014). "*Financial Times* Top 45 Journals Used in Business School Research Rankings." http://www.ft.com/cms/s/2/3405a512-5cbb-11e1-8f1f-00144feabdc0.html#axzz3C7N5dkNJ (accessed September 2, 2014).

Gawande, A. (2010). "The Checklist. If Something So Simple Can Transform Intensive Care, What Else Can It Do?" http://www.newyorker.com/magazine/2007/12/10/the-checklist (accessed September 2, 2014).

Gawande, A. (2011). "Personal Best. Top Athletes and Singers Have Coaches. Should You? http://www.gopaldas.net/uploads/2/5/1/2/25121492/gawande_2011_the_new_yorker.pdf (accessed September 2, 2014).

Gilbert, E. (2006). *Eat, Pray, Love: One Woman's Search for Everything Across Italy, India and Indonesia.* New York: Viking Press.

Goffman, E. (1959). *The Presentation of Self in Everyday Life.* New York: Anchor.

Harzing, A.-W. (2014). "Journal Quality List. Fifty-Second Edition, 11 February 2014." http://www.harzing.com/download/jql_journal.pdf (accessed, September 2, 2014).

Keynes, J.M. (1936). *The General Theory of Employment, Interest and Money.* http://cas.umkc.edu/economics/people/facultyPages/kregel/courses/econ645/Winter2011/GeneralTheory.pdf (accessed September 7, 2014).

Uncles, M.D. and Kwok, S. (2013). Designing research with in-built differentiated replication. *Journal of Business Research,* 66: 1398–1405.

Üstüner, T. and Thompson, C.J. (2012). How marketplace performances produce interdependent status games and contested forms of symbolic capital. *Journal of Consumer Research*, 38: 379–381.

Urry, J. and Larsen, J. (2011). *The Tourist Gaze 3.0.* Thousand Oaks, CA: Sage.

Whyte, W.F. (1943). *Street Corner Society: The Social Structure of an Italian Slum.* Chicago: University of Chicago Press.

Woodside, A.G. (2014). Embrace•perform•model: Complexity theory, contrarian case analysis, multiple realities. *Journal of Business Research*, 67: 2495–2503.

3

WELL-DONE LITERATURE REVIEWS

A Journal's Editor-in-Chief Perspective

MURRAY E. JENNEX

Contents

Introduction

As academics and researchers, we are taught to conduct and report on our research. One of the basic skills we learn is to conduct and write the literature review. Jennex (2009) discusses the value of good literature reviews as being the building and strengthening of a body of knowledge. Good research that is not basic research builds on that which was done before and uses previous research to ground current research in theory and as a lens for interpreting results. However, as the numbers of information systems (IS) journals grow—Lamp (2004) lists 861 IS journals with 735 still active (as of March 25, 2014)—the time and effort it takes to conduct a thorough and comprehensive literature review is growing. The result of increasing time to conduct and report on the literature review is that the quality of the literature review is declining. Of course, fueling the growth in IS journals is the increase in academics and researchers conducting and submitting research. The result of having more research articles being written is that there are more to review, creating a drain on limited reviewer resources. The data to support these statements comes from reviewing the reviews of articles and papers submitted to the *International Journal of Knowledge Management* (*IJKM*). Other experience (anecdotal influencing my perception) is used from reviews from the *International Journal of Information Systems for Crisis Response and Management* (*IJISCRAM*), the Hawaii International Conference on System Sciences (HICSS), the Association of Information Systems (AIS) Conference of the Americas (AMCIS), European Conference on Information Systems (ECIS), and various other ad hoc reviews for IS journals and doctoral dissertations. This chapter first discusses how literature adds value to research. Next the chapter discusses reasons offered or observed that are used to justify lower-quality literature reviews. The chapter concludes with suggestions on what should be done to improve literature review quality and perhaps establish a baseline publishing policy on literature reviews.

Why Are Literature Reviews a Problem?

Why is this trend of decreasing literature review quality with increasing numbers of articles to review an important issue? Our IS discipline, as is every other discipline, is very concerned with plagiarism be it intentional or accidental. Postings on plagiarism periodically appear on our discipline's list server, ISWorld. The discipline is very concerned that authors may publish others' work as their own. Journals have or are adopting tools and guidelines to assist them in ensuring that articles are sufficiently different from previously published work. AIS has published and used a procedure for authors to grieve and have resolved charges of plagiarism. Good literature reviews help authors avoid accidental plagiarism by ensuring they have documented what is known and given credit where it is deserved. Additionally, authors have to be careful of self-plagiarism, or in other words, of copying their own work without attributing proper credit to their work. A special issue of the *Communications of the AIS* that focused on citation issues including self-citation and recommended citations from reviewers and editors has been published. Ultimately a good quality literature review prevents plagiarism from accidentally happening.

Also, the value of our research is how it contributes to what Jennex (2009) calls the body of knowledge. The value of the contribution of research is measured by the number of citations a paper and an author receive. Citation counts for authors, as measured by the h index, are an important measure of the relevancy and importance of our work. Seminal papers are determined by how many cite a paper. The issue of declining quality of literature reviews directly affects these measures and as such should be a big concern for all researchers.

Finally, we are seeing rapid growth in the numbers of articles to review caused by growing numbers of researchers from universities in Asia and the Middle East contributing research to the traditional Western-based IS journals. Additionally, we are finally seeing growth in the numbers of junior faculty as the hiring impact from the 2008–2009 economic crisis subsides, and new faculty are being hired. Of course, these new academics and researchers are feeling the pressure of publish or perish and are more concerned with conducting and publishing their research rather than review others' research. Reviewing has traditionally been the service provided by senior academics and researchers. There is starting to be a decline in the numbers of senior faculty available to provide this service and it is expected that this will be a growing trend over the next several years as the "baby boomer" generation begins to retire (Jennex, 2013). The net result of more junior faculty and fewer senior faculty is that there is a decreasing pool of willing reviewers, making reviewer time a limited commodity that needs to be managed. Is it more important for reviewer resources to be spent on assessing the literature review or should they focus on reviewing the actual research? My position as a journal editor-in-chief is that I want to spend these resources on reviewing the research, not reviewing literature reviews.

Ultimately, the issue this chapter is addressing is the impact that lower-quality literature reviews is having on our limited reviewer resources and what can be done to address it and thus maximize the value of these limited reviewer resources.

Methodology

The methodology for identifying the literature review as a problem is fairly simple. Numerical summations of numbers of papers with and without comments on the literature review are used from the *International Journal of Knowledge Management*. This information is gathered through the author's role as editor-in-chief of the journal. The author is also editor-in-chief of the *International Journal of Information System for Crisis Response and Management* (*IJISCRAM*) and is a track chair at HICSS, minitrack chair at AMCIS, associate editor at *ECIS*, and ad hoc journal reviewer and external doctoral dissertation examiner. These other sources are not used for various reasons. First is that *IJISCRAM* is a journal for a new field and Webster and Watson's (2002) comment referring to the slow growth of theory is particularly true for this journal. All submissions suffered literature review comments as the crisis response community is fragmented and spread across many disciplines. ECIS and AMCIS data are not presented inasmuch as the numbers of papers where review comments are available are low, making it too easy to link to specific authors. Also, the numbers of doctoral dissertations where review comments are available is very low, again making it very easy to link data to specific authors. No author, university affiliation, or other identifiable information is provided. This chapter is not meant to embarrass or single out any individual nor make any accusations. The only purpose of this chapter is to discuss what I perceive to be a disturbing trend in reporting on IS research.

The methodology for generating the literature review was to use Google Scholar and the AIS e-library search engines to search the literature reviews, and conduct research. The purpose of the literature review was to look at the Internet and the AIS e-library, respectively. Three search terms were used; articles prior to 1990 were discarded. Also, this chapter is focused on IS research discussions looking at literature reviews in other disciplines/fields and differences between IS and other disciplines/fields were not incorporated.

Literature Review

Lamb (2013) defines the literature review as a review of secondary sources documented in text that considers the critical points of current knowledge including substantive findings, as well as theoretical and methodological contributions to a particular topic. A systematic review is a literature review that addresses a research question by identifying, appraising, selecting, and synthesizing all high-quality research evidence relevant to that question (also Fink, 2005). The key words are "all high quality." "All" implies a wide-ranging search that is difficult when we remember that there are 861 journals that need to be checked, plus perhaps journals from other related disciplines. High quality is difficult to define. Does high quality only refer to the top tier of journals, and if so, what are the top tier journals? This is a subjective call. Do we only consider the AIS Senior Scholar basket of journals to be the only sources of high-quality research? I would hope not, given that there are 861 IS journals but only 8 listed in the basket, implying that only approximately 1% of our journals publish high-quality research.

The University of Arizona (2011) describes the two purposes of the literature review. The first is to justify your research by showing there are gaps of knowledge that are worthy of closer investigation, that the contribution is original, that the research has been approached in a rigorous manner, and to show if existing research contradicts or supports the research approach. The second is to develop the thesis position by showing understanding of the critical literature, identifying issues, and framing the research into what is known, what remains to be learned, and how the research will contribute. Dennis and Valacich (2001) summarize the purpose of the literature review to identify theory that can be used to explain findings and to conduct the research. Dennis and Valacich (2001) also identify as the top two ways of getting rejected by a quality journal as avoiding theory in favor of a summary of prior research and omitting key papers from your literature review. The conclusion from these sources is that the literature review is more than just a summary of the literature; it also frames the research in theory. This is important to understand as it suggests that the literature review is a very important part of research and not just something that we are required to do.

Fink (2012; note that there are previous Fink citations on related work starting in 1998) discusses the process of doing a systematic literature review and breaks the systematic literature review into seven tasks (note that these tasks also correspond to what is commonly called a literature review; thus to avoid confusion the rest of this chapter uses the term "literature review" also to refer to a systematic literature review):

1. Selecting research questions
2. Selecting bibliographic or article databases
3. Selecting search criteria
4. Applying practical screening criteria
5. Applying methodological screening criteria
6. Doing the review
7. Synthesizing the results

These steps have been further enhanced by Brocke et al. (2009) who stated that the quality of a literature review depends upon the rigor of the search process, that is, steps 2 through 5 above. In addition, Bandara, Miskon, and Fielt (2011) propose that the quality of a systematic literature review is improved by using tools such as Google Scholar, Endnotes, and the like to aid in identifying appropriate articles and NVIVO for coding, interpreting, and synthesizing literature. Finally, Davison, de Vreede, and Briggs (2005) state that it is the duty of the reviewer to ensure the quality of the literature review by ensuring appropriate citations and theory are used in the papers under their review.

Limitations to the above literature review process come from a few sources. Boell and Cezec-Kecmanovic (2011) contend that there is little difference between systematic literature reviews and nonsystematic literature reviews (hence why this chapter just refers to the term "literature review"). To improve overall literature review quality, Boell and Cecez-Kecmanovic (2014) propose a hermeneutic approach to understanding the literature in literature reviews and provide several steps for researchers to understand and synthesize the literature. Sammon et al. (2011) recognize that doctoral students and new researchers have a difficult time understanding and synthesizing theory in the literature review and propose that these researchers focus on creating a pedagogical artifact. Webster and Watson (2002) lamented that theory-building progress in IS was slow due to a lack of review articles and the newness of the field. Finally, Okoli and Schabram (2010) use the Fink (2012) steps for conducting the literature review but recognize the difficulty in doing them well and so provide a focus on the practical screen. The practical screen is the process used to narrow down the articles to use in the literature review. They identify the following as acceptable reasons to drop articles from consideration:

- *Content (topics or variables):* The review must always be practically limited to studies that have bearing on its specific research question.
- *Publication language:* Reviewers can only review studies written in languages they can read, or for which they have access to scholarly databases.
- *Journals:* The scope of the review might limit itself to a select set of high-quality journals, or include only journals in a particularly field of study.
- *Authors:* The study might be restricted to works by certain prominent or key authors (potentially including the reviewer).
- *Setting:* Perhaps only studies conducted in certain settings, such as specific industries or regions, might be considered.
- *Participants or subjects:* Studies may be restricted to those that study subjects of a certain gender, work situation, age, or other pertinent criteria.
- *Program or intervention:* There might be a distinction made between the nature of the measurement in the studies, such as if data is self-reported versus researcher-measured, or if subjects are self-selected into various groups within the study.
- *Research design or sampling methodology:* Studies might be excluded based on not using a particular research design. Note that there are significant differences between these judgments between disciplines.
- *Date of publication or of data collection, or duration of data collection:* Studies will often be restricted to certain date ranges.
- *Source of financial support:* Studies might be restricted to those receiving nonprivate funds unless there is a concern that this might be a source of bias in the results.

The validity of the above set of reasons for limiting a literature review is discussed later in this chapter.

A Google search on conducting literature reviews found that there are many universities with online guides for conducting literature reviews. Two were reviewed, the Writing Center of the University of North Carolina and the Writing Handbook of the University of Wisconsin both contain guidance that is similar to what has been discussed and show that there is some consistency in the teaching of young researchers as to how to perform the literature review.

To summarize and synthesize this section it is found that for IS research we are teaching researchers to perform systematic literature reviews (or simply the literature review). The purpose of this review is to summarize the relevant literature and to synthesize theory to frame results and to provide research approaches and measures. Additionally, the literature review shows that the paper is making an original contribution and gives credit to where credit is due.

Issues with the Literature Review

The primary data to support the statement that there is a literature review quality problem comes from an analysis of review comments on submissions reviewed from January 1, 2013 through March 1, 2014. A total of 81 submissions were reviewed (submissions currently in review are not included in this number) during this period with the following distribution of significant review comments on the literature review. Note that for this analysis a significant literature review comment is one where more than five sources are recommended or where entire topic areas or journals are recommended.

The knowledge management discipline is a small but growing discipline with a core of KM-focused journals. Serenko and Bontis (2009) ranked KM and innovation journals and listed 17 KM-focused journals. Serenko and Bontis (2013) updated the 2009 paper and listed 21 KM-focused journals or approximately 3% of the active journals listed by Lamp (2004). Also, KM papers have been published in many other journals including journals listed in the AIS Senior Scholar basket of journals. In addition, there are several conferences such as HICSS, AMCIS, and ECIS that have dedicated tracks or minitracks for KM. It can be summarized that the body of knowledge from which to conduct KM literature reviews is small compared to the overall IS body of knowledge.

Table 3.1 indicates that approximately 70% of submissions had significant literature review comments. Although data is not provided, a 70% rate of significant literature review comments is a good heuristic for what was seen in the conference submissions. All the ad hoc reviews and the doctoral dissertations reviewed had significant

Table 3.1   Submissions with Significant Literature Review Comments

PAPER DISPOSITION	NUMBER WITH SIGNIFICANT LITERATURE REVIEW COMMENTS	NUMBER WITHOUT SIGNIFICANT LITERATURE REVIEW COMMENTS
Accept/Conditional Accept[a]	5	11
Revise	49	7
Reject	3	6
Totals	**57**	**24**

[a] Note that these issues were fixed prior to publication.

literature review comments. As an editor-in-chief, I consider a 70% significant comment rate on literature reviews to be unacceptable as it has impacts on the ability of reviewers to do a quality review as well as on the citations necessary to build the KM body of knowledge.

The high incidence of significant literature review comments has an impact on the workload of the reviewers as Davison, de Vreede, and Briggs (2005) state that reviewers need to provide specific citations/references as a part of the review. This chapter defines a significant literature review comment as either missing five or more references or missing references from entire journals or topics implying that the reviewer workload to meet the expectation of providing specific references is going to include performing all or part of the literature for 70% of the submissions. There are few senior researchers willing to do all or part of the literature review for an author on a review at *IJKM*; it is imagined that this is true at most if not all journals. The result is that as editor-in-chief I reject or require major revision on articles without providing the quality review as described by Davison, de Vreede, and Briggs (2005) and so authors are taking longer to complete their research. In this age of Internet publishing and expectations of 30-day review cycles, can journals afford to require this amount of time and effort? I would prefer authors put this effort in initially so that reviewers can focus on the merits of the research.

The next question to consider is why the literature reviews warrant significant comments. The following are the observed reasons deduced from the reviewer comments:

1. *Literature reviews of convenience:* These literature reviews are usually done by authors who do not have immediate access to all the relevant KM articles. This was the most commonly

observed reason for a poor literature review and is detected when the article's literature review contains KM articles from only one to a few of the KM journals, usually the open access journals or those journals available through EBSCO or other online repositories. The most common response to this comment by authors is that our university cannot pay for access.

2. *Weak search criteria:* These literature reviews are usually done by authors who want to ensure that they are doing new work. This was the most commonly observed reason for a poor literature review by students and new or very junior academics/researchers and is detected by search criteria that are not consistent with the logical breakdown of the subject being examined (examples include using partial ontology such as knowledge transfer and not associated terms such as knowledge flow or knowledge sharing; or using new names for constructs that already have agreed-upon ontology such as a knowledge management repository system rather than the common term knowledge management system).

3. *Artificial search criteria:* These literature reviews are usually done by authors who want to limit the number of articles they need to include in the literature review. This was observed in several cases with no discernable pattern for its use. These literature reviews are characterized by constrained search criteria (examples include search criteria that only look at journals in the AIS Senior Scholar basket or search criteria that are regionally constrained such as search criteria that only look for articles in South Africa or articles written in a language other than the language of the journal, such as Chinese, or search criteria that only look at articles identified as quantitative and ignore qualitative articles that use quantitative measures).

4. *Not going to the source:* These literature reviews are done by authors who may not know better or who do not have access to the source articles. This is rapidly becoming a major issue due to the open source movement and Internet postings. This is detected by reviewers who understand the seminal works

and key concepts. It is characterized by literature reviews that cite an article that cites an article. Instead of following up to the source document, the authors are content to cite the document they are reviewing as the source document. (An example would be citing a Jennex (xxxx) article for a point made by Alavi and Leidner (xxxx) because the author has the Jennex article but not the Alavi and Leidner article.) This issue is potentially the most damaging as it causes authors not to build on the existing body of knowledge and can potentially damage colleagues by not giving the appropriate credit where it is due. The issue is becoming more prevalent due to authors citing Wikipedia instead of the source citation; authors citing an edited book author instead of the chapter author; or authors who cite the open source article instead of the cited source in the document. It is suspected that this could also be an issue with journals in other languages due to translation errors or lack of knowledge on how to cite properly by the party performing the translation.

5. *Not understanding the source:* These literature reviews usually do a good job of summarizing the literature but fail to synthesize it or even worse, incorrectly synthesize the knowledge in the source. Reasons for these literature misinterpretations vary and many may be due to translation issues for nonnative English speakers.

Of course there are other reasons for significant literature review comments but the above five account for the vast majority.

Recommendations

What can be done to improve the quality of literature reviews? It is not the position of this chapter that we are not training academics and researchers correctly. It is the position of this chapter that the acceptable reasons for limiting literature reviews using the practical screen (Okoli and Schabram, 2010) are being incorrectly applied and perhaps should be done away with. All five of the main reasons for low-quality literature reviews can be justified by the practical screen criteria of Okoli and Schabram (2010). It is not the position of this chapter that Okoli and Schabram (2010) intentionally encouraged poor-quality literature reviews. This chapter suggests that authors have used the practical screen to justify poor-quality literature reviews. This chapter makes the following recommendations:

For authors:

- Understand that the purpose of the literature review is to ground the research in the literature as well as to recognize what has been done and what has not. This includes using validated instruments and using the theory to explain your results.
- Heed the advice of Dennis and Valacich (2001) and ensure that you have synthesized the literature as well as summarized it and that you understand and know the critical papers in the discipline; never use a citation of a citation.
- Include in the methodology section the methodology used to perform the literature review. Include in this discussion search criteria, repositories searched, and reasons for excluding articles.
- Utilize all the search tools and repositories at your disposal per Bandara, Miskon, and Fielt (2011).
- Know what the journals are in your field and be sure to search all of them. If you do not have access to an article through your university, contact the authors directly or at least search repositories such as ResearchGate to see if you can get access to the article. If the article is highly relevant to the research then do what it takes to get it!
- Never settle for "enough" articles in your literature review. There is no minimum and the right number of articles are all those that are relevant to the research.
- Understand the ontology of the field/discipline and use multiple search criteria from this ontology to search for relevant articles. Do not create your own ontology for the field or discipline as this displays a lack of understanding and respect for the field/discipline.
- Do not ignore literature that contradicts your research or that suggests doing something different from what you did.

In particular if you find literature after the fact that suggests constructs or instruments you did not use, do not ignore it. Include it in some manner, perhaps as future research, limitations to research, or in the discussion on what the research means.

- Make a good faith effort to review all the relevant literature.

For editors:

- Do not burden your reviewers with reviewing unacceptable or very low-quality literature reviews: desk-reject the paper.
- Assist authors in finding appropriate literature from your journal and encourage reviewers to suggest their own articles if they are relevant as suggested by Jennex (2009).
- Be aware of the journals in your field or discipline so that you can ensure authors are covering them.
- Do not accept the reasons of the practical screen (Okoli and Schabram, 2010) for limiting literature reviews.
- Include guidance, expectations, and best practice for performing literature reviews in the guide to authors.

For reviewers:

- It is not your job to perform the literature review for the author and it is OK to tell the author to do his or her job.
- Understand the ontology of the field or discipline and ensure that the methodology used to perform the literature review is appropriate.
- Recommend a strategy for doing the literature review when there are significant issues with the literature review.
- Ensure the critical papers in field or discipline are reviewed as appropriate (Dennis and Valacich, 2001).
- Ensure that authors synthesize the literature and demonstrate correct understanding of the literature; expect more than a summary of articles (Dennis and Valacich, 2001)!
- Do not consider the literature review as just something that needs to be done; it is an important part of research and ultimately the goal is to further the body of knowledge.

Conclusions

This chapter posits that there is a drastic decrease in the quality of literature reviews being included in submitted research. The reasons are many but are probably due to the large number of IS journals (861 per Lamp, 2004) and may all be based on the acceptable reasons for limiting a literature review, or the practical screen (Okoli and Schabram, 2010). The result of lower-quality literature reviews is increased workload on reviewers at a time when there are more papers being submitted to journals and conferences. There is no simple fix to the problem. The main suggestions are for authors to include a methodology used to conduct the literature review in the methodology section of the paper; editors to include guidance, standards, and expectations for the performance of the literature review in the guide to authors; and reviewers to suggest a strategy for performing the literature review rather than focusing on recommending specific papers. Although it will be difficult to improve the quality of literature reviews, it is critical that we do so to ensure that credit is given where deserved and in the process build on the body of knowledge and so that we can manage the workload we are expecting of our limited reviewer resource.

References

Bandara, W., Miskon, S., and Fielt, E. (2011). A systematic, tool-supported method for conducting literature reviews in information systems. In *ECIS 2011 Proceedings*, Paper 221.

Boell, S. and Cezec-Kecmanovic, D. (2011). Are systematic reviews better, less biased and of higher quality? In *ECIS 2011 Proceedings*, Paper 223.

Boell, S.K. and Cecez-Kecmanovic, D. (2014). A hermeneutic approach for conducting literature reviews and literature searches, *Communications of the Association for Information Systems*, 34, Article 12.

Brocke, J.v., Simons, A. Niehaves, B., Niehaves, B., Reimer, K., Plattfaut, R., and Cleven, A. (2009). Reconstructing the giant: On the importance of rigour in documenting nthe literature search process. In *ECIS 2009 Proceedings*, Paper 372.

Davison, R.M., de Vreede, G.J., and Briggs, R.O. (2005). On peer review standards for the information systems literature. *Communications of the Association for Information Systems*, 16, Article 49.

Dennis, A.R. and Valacich, J.S. (2001). Conducting research in information systems. *Communications of the Association of Information Systems*, 7, Article 5.

Fink, A. (2012). *Conducting Research Literature Reviews*. Thousand Oaks, CA: Sage.

Fink, A. (2005). *Conducting Research Literature Reviews: From the Internet to Paper*, 2nd ed. Thousand Oaks, CA: Sage.

Jennex, M.E. (2013). Knowledge management: The risk of forgetting. *iKNOW, the Magazine for Knowledge Workers*, 3(1): 4–7.

Jennex, M.E. (2009). Journal self citation VII: Building a body of knowledge. *Communications of the Association of Information Systems*, 25, Article 7: 67–72.

Lamb, D. (2013). "The Uses of Analysis: Rhetorical Analysis, Article Analysis, and the Literature Review." *Academic Writing Tutor*. Retrieved March 18, 2014 from http://www.academicwritingtutor.com/uses-analysis-rhetorical-analysis-article-analysis-literature-review/

Lamp, J.W. (2004). *Index of Information Systems Journals*. Geelong, Deakin University. [Online] Available: http://lamp.infosys.deakin.edu.au/journals/

Okoli, C. and Schabram, K. (2010). A guide to conducting a systematic literature review of information systems research. *All Sprouts Content*. Paper 348.

Sammon, D., Nagle, T., O'Raghallaigh, P., and Finnegan, P. (2011). Design of a pedagogical artefact for doctoral researchers to assess theoretical strength. In *ECIS 2011 Proceedings*. Paper 232.

Serenko, A. and Bontis, N. (2013). Global ranking of knowledge management and intellectual capital academic journals: 2013 update. *Journal of Knowledge Management*, 17(2): 307–326.

Serenko, A. and Bontis, N. (2009). Global ranking of knowledge management and intellectual capital academic journals. *Journal of Knowledge Management*, 13(1): 4–15.

The Writing Center. (2014). "Literature Reviews," University of North Carolina Writing Center. Retrieved March 19, 2014 from https://writingcenter.unc.edu/handouts/literature-reviews/

The Writing Handbook. (2014). "Learn how to write a review of literature," The University of Wisconsin Writing Center. Retrieved March 19, 2014 from http://writing.wisc.edu/Handbook/ReviewofLiterature.html

University of Arizona. (2011). "The Literature Review—Purpose." Retrieved March 28, 2014 from http://www.library.arizona.edu/help/tutorials/lit-reviews/whatis.html

Webster, J. and Watson, R.T. (2002). Analyzing the past to prepare for the future: Writing a literature review. *MIS Quarterly*, 26(2): xiii–xxiii.

4

POSITIONING YOUR PAPER FOR PUBLICATION IN A JOURNAL

Where Do Authors Go Wrong?

ALEX KOOHANG

Contents

Introduction

This chapter discusses some of the most common mistakes authors make in positioning their papers for submission to a journal for possible publication. Methods of minimizing these mistakes are discussed. Authors are provided with useful suggestions to make their papers suitable enough to (1) pass the screening process by the screening committee so it will advance to the review process and (2) make it easier for reviewers to review and provide their assessment/feedback to authors.

"Communication Platform"

The "Communication Platform" in a paper submission system is the main element that connects authors to the editor of a journal.[*] The five major tasks of a communication platform of a paper submission system are:

1. Author Submission Task—Author electronically submits his/ her manuscript for possible review and possible publication in the journal.
2. Initial Editorial Task—The Editor screens the manuscript to determine its suitability for possible publication in the journal. If the manuscript is determined suitable for possible publication, the Editor assigns it to two or more appropriate reviewers who are comfortable reviewing the paper.
3. Reviewers Task—The reviewers are asked to review the manuscript and make recommendation regarding acceptance or rejection of the manuscript. If the reviewer(s) lean toward acceptance, they are asked to provide constructive feedback on how to improve the paper for revisions. If the paper is rejected, the reviewer is also asked to provide positive feedback on how the paper can be improved.
4. Second Editorial Task—Once the reviews are completed, the Editor communicates with the author and provides detailed

---

[*] Koohang, A. and Harman, K. (2006). The academic open access e-Journal: Platform and portal. *Informing Science Journal*, 9: 71–81.

feedback regarding the decision for publication. The Editor will do the same for papers that are rejected.

5. Revision Task—The author revises the manuscript and re-uploads the manuscript into the system.
6. Third Editorial Task—The Editor then confirms the revisions by sending the paper for further review, and if the paper meets the acceptance criteria, the Editor communicates with the author and prepares the paper for the Content Management Platform. *

The above tasks are quite common to most communication platforms of paper submission systems. In the case of the *Journal of Computer Information Systems* (*JCIS*), where this writer serves as its editor-in-chief, the tasks are at four levels. These levels are (1) review by the screening committee, (2) reviewer assignment, (3) the preliminary decision, and (4) the final decision. The rest of this chapter delineates each of these levels of the communication platform, points out the common mistakes that are made by authors in each level, and provides useful suggestions as to how these mistakes can be eliminated to improve the chance of the papers being published.

Level 1: Review by the Screening Committee

Level 1 of the communication platform is the review of each submitted paper by the screening committee. The screening committee consists of the editor-in-chief and two persons on the editorial board. All submitted papers are examined by the screening committee. The screening committee checks for those manuscripts relevant to (1) the interest to the *JCIS* readership; (2) the fit with the journal's aims and scope; (3) the large number of same topic submissions in a given period of time; or (4) compliance with the author's guidelines and policies. The screening committee will then decide whether the manuscript advances to Level 2 of the communication platform. If the decision is not to advance to the next level, the authors are immediately notified.

One of the major elements influencing the decision of rejecting the paper at Level 1 is the fact that the authors do not follow the guidelines

---

* Koohang, A. and Harman, K. (2006). The academic open access e-Journal: Platform and portal. *Informing Science Journal*, 9: 71–81.

and policies of the journal. In this section, the manuscript submission policies and guidelines for the *Journal of Computer Information Systems* are used as an example to outline common mistakes authors make (when submitting their papers for possible publication) that may inhibit the advancement of their papers to Level 2: Reviewer Assignment. These policies and guidelines are directly taken from the *Journal*'s website.* In the next part, each of these policies and guidelines is discussed. Common mistakes are pointed out and suggestions are made for improvement.

Policies

The Submitted File Format Is Wrong JCIS: "Electronic Submission: Authors submit their manuscripts (in MS Word Format) electronically using the JCIS Paper Submission and Review System."

When submitting a manuscript for possible publication in a journal, authors must make every effort to follow the guidelines for the required file format of the journal. Although this is not a frequent mistake based on this writer's experience, the authors must check the file format and adhere to the journal's guidelines. Some paper submission systems check for the file format and alert the author. Keep in mind that not all systems do this. You do not want to submit a paper in a file format that cannot be opened by the editor.

Authors Submit Partially Published Work or Work under Consideration JCIS: "Originality: All manuscripts must be the authors' original, unpublished work. The manuscript must not be under consideration for publication elsewhere."

Authors have an ethical obligation to adhere to this policy. Originality means that the paper is the author's original work. It has not been published in another outlet. It is not under review by another journal. And, it is not someone else's work. For this reason, journals require all authors to adhere to the "research code of conduct" that outlines ethical obligations of authors conducting research.†

---

* http://www.iacis.org/jcis/guidelines.php
† See AIS Research Code of Conduct at http://aisnet.org/

Also note that increasingly, journals are placing a plagiarism detection system as part of the communication platform of the paper submission system to check for paper originality and possible plagiarism.

Occasionally, authors submit papers to journals for possible publication that are published in the conference proceedings. Some journals may accept these papers for possible publication if considerable changes are made from the earlier versions. In such cases, the authors must acknowledge the earlier version and source of publication in the new version of the paper. This must also be noted in the cover letter to the editor.

Infringing Copyright Materials   *JCIS*: "Copyright: The paper does not infringe the copyright, or violate other rights, of any third party."

The authors must make sure that the paper does not infringe any copyright. Copyright infringement means that "a copyrighted work is reproduced, distributed, performed, publicly displayed, or made into a derivative work without the permission of the copyright owner." *

Journals often require that authors obtain permissions to use reprinted material, adapted material, and material owned by other parties.

Leaving the Author(s) Information on the Submitted Paper *JCIS*: "Blinding Manuscript Submission: Authors(s) must remove any identifying information (name, affiliation, etc.) from the paper prior to submission. Information regarding authors is collected by the system on the manuscript submission form. Authors should not indicate on their résumé or website that a paper is under review at *JCIS*."

For a peer review process, a paper must not include authors' identifying information. They must be removed before submitting a paper for possible peer review and possible publication in a journal. This is one of the careless mistakes authors make and must be avoided.

In addition, there are three important tasks that must be followed. First, the authors must remove any copy/version of the submitted paper from any public websites as it may interfere with the blind review. Second, if the author is citing his or her prior research, the appropriate practice for citing the work in the body of the text is as follows:

---

* http://www.copyright.gov/help/faq/faq-definitions.html

Author (2010) or (Author, 2010).

Author and coauthor (2012) or (Author and coauthor, 2012)

In the reference section, it should then read:

Author (2010). Deleted for peer review.

Author and coauthor (2004). Deleted for peer review.

Third, once your paper is accepted by the screening committee for advancement to the next level (assignment to two reviewers) the authors must not indicate on their résumé or website that the paper is under review.

Authors Ignore the Journal's Instructions/Guidelines Posted on the Journal's Website *JCIS*: "Note: *JCIS* does not accept letters of inquiry, abstracts, or topic proposals for pre-evaluation of papers to determine their suitability for possible review/publication. Authors must review the *JCIS* suggested topics available at http://www.iacis.org/jcis/jci_cfp/ JCIS_CFP.pdf and submit their finished work for possible review/ publication to the *JCIS* Paper and Review Submission System."

The above policy may not be a universal policy for all journals; however, it is a policy of the *JCIS*. Often, the editor-in-chief receives letters of inquiry, topic proposals, abstracts, or (sometimes the completed paper) for pre-evaluation by the editor before they submit their papers. The authors often request an answer about the suitability of their papers for the journal.

Frequently, this is the result of the authors' not reading the policy or ignoring the policy. Journals with such a policy have clear directions/ instructions on where (and how) the authors can find the information. Such inquiries are replied to by the editor-in-chief, pointing the authors to the manuscript submission policies page that includes the information they have requested.

Manuscript Guidelines

Overlooking the Manuscript's Guidelines Ignoring a journal manuscript's guidelines is one of the major mistakes authors make when submitting their manuscripts for possible review and publication. Manuscript guidelines often refer to a journal's specific rules and practice in which a paper should be formatted. Each journal has its own formatting guidelines. The most common guidelines include information about the manuscript length (i.e., the maximum number of pages or words allowed); page layout (i.e., the margins, columns, fonts, indentation, spacing); maximum number of words used in the title of the manuscript; headings and subheadings format; layouts of the tables and figures; maximum number of words used in the abstract section; a set of keywords as the indicators to the content of the manuscript; and the style in which the references should be used in the manuscript (i.e., APA, MLA, Harvard, Chicago, Vancouver). Below is a list of common mistakes that authors make with their manuscripts:

- Manuscript length far exceeds the maximum number of pages or words allowed by the journal.
- Use of fonts (size or theme) other than what the journal allows.
- The manuscript uses line spacing other than what the journal requires (i.e., single-spaced for double-spaced or vice versa).
- The title exceeds the number of words required by the journal.
- The journal's guideline for formatting the headings and subheadings is not followed (i.e., use of different font size or theme, numbering the headings when the guideline specifically indicates otherwise).
- The journal's guideline for formatting (and numbering) of the tables and figures is not followed.
- The abstract length far exceeds the number of words allowed by the journal. Some journals require a structured abstract with specific headings, that is, purpose, design/methodology/approach, findings, research limitations/implications, practical implications, originality/value. (See *Industrial Management & Data Systems Journal*.)
- Keywords are missing or required number of keywords is wrong.
- The style in which the references should be used in the manuscript is different than what the journal requires (both cited within the text and in the reference section).

Occasionally, an author submits a paper to a journal that has already been rejected by another journal. This is a common practice; however, the author must comply with the new journal's manuscript guidelines. Sometimes, papers are submitted exactly with the old journal's manuscript guidelines (the rejecting journal) to the new journal. This is a bad mistake that authors should avoid.

Often, the rejecting journal provides authors with a set of reviewers' comments/feedback. Authors must use these comments/feedback to improve their paper and submit to another journal keeping in mind to follow the new journal's manuscript guidelines carefully.

Level 2: Reviewer Assignment

Level 2 of the communication platform is the reviewer assignment task. The manuscripts passed by the screening committee are assigned by the editor-in-chief to two or more reviewers who are knowledgeable in the topic. Reviewers are given 30 days to complete the review. Reviewers are asked to assess the article and go through two review steps.

Step 1

Choose one of the four categories below:

Category 1: Paper is weak in all the following areas: relevance, originality/novelty, importance, and clarity. (Please complete Step 2.)

Category 2: Paper is not as badly flawed, but a major effort is necessary to improve the paper. (Please complete Step 2.)

Category 3: Paper has merit, but accuracy, clarity, completeness, and/or writing should be improved. (Please complete Step 2.)

Category 4: Paper meets professional norms: relevance, originality/novelty, importance, and clarity. (Please complete Step 2.)

Step 2

In Step 2, reviewers are required to provide a narrative assessment containing their candid and fair assessment of the article based on the category they have chosen in Step/Level 1. In assessing the paper, the reviewers are asked to analyze the strengths and weaknesses of the paper and provide suggestions for improvement. In the next part, the most common mistakes made by the authors are described.

Missing Text about How the Paper Is Organized   Authors often forget to include a paragraph describing the organization of their paper. Authors must write a paragraph—normally in the beginning of the paper—describing how their paper is organized. This offers the reviewers a sense of what to expect.

The Study Design   The overall study design and organization of a manuscript is very crucial in the review process. A sound study design will begin with the statement of the problem. In other words, what is the problem being studied/researched? The statement of the problem will follow the study's purpose. Consistent with the purpose of the study, the research questions or hypotheses should be stated. The variables included in the study must be defined and justified.

Subsequently, a sound review of the literature follows. The literature review must be relevant and consistent with the problem being examined. The review of the literature is followed by the research/study methodology. The authors must clearly state the reliability and validity of the instrument used in the study to collect data for analyses (and to answer the research questions/hypotheses). If an instrument is not tested for reliability and validity, the study can be considered flawed. Next in line, a reviewer will look for the following:

- Are the participants and the population sample well-defined? If a convenience sample is used, is it justified to represent the population?
- Was the proper procedure followed? How was the study administered, where was it administered, and was the approval to administer the study to subjects obtained from the Institutional Review Board? Was anonymity guaranteed to the participants?
- Are appropriate data analyses used and justified, along with underlying stated assumptions?
- Are the results of the study clearly stated?
- Does the paper include a discussion of findings? Is the discussion of findings supported (or not supported) by the previous research?
- Are the implications of the study clearly described?
- Are the recommendations for future research stated?
- Are the limitations of the study clearly communicated?

The paper must also indicate the importance, the originality, and the value it is providing to the field of research (i.e., what is the significant contribution of this research versus what already exists in the published literature). Once the reviews are completed, the editor-in-chief moves into Level 3.

Level 3: The Preliminary Decision

Level 3 of the communication platform is the preliminary decision. The editor-in-chief reviews and evaluates the reviewers' narrative assessments and makes a final decision whether to reject the manuscript or conditionally accept the paper pending revision and final formatting. If the paper is rejected, then a detailed e-mail is sent to the author with the reviewers' narrative assessments.

If the paper is conditionally accepted, then the author is asked to revise the paper (in 25 days or less) to address the reviewers' narrative assessments. The author will be asked to provide a separate document that lists how he or she addressed (or was unable to address) each of the reviewers' narrative assessments. To speed up the process of final publication, the author is asked to comply carefully and fully with the journal's guidelines and requirements, and provide a separate page with the following information:

1. Title, no acronyms, not to exceed 10 words.
2. Authors' names in the order of first submission (an asterisk should be next to the corresponding author), institutional affiliation, address, phone number, fax number.
3. E-mail address(es) separated by comma for each author.

The author will then send the files to the editor-in-chief. The editor-in-chief reviews the revised manuscript and determines whether it requires a second round of revision. The most common mistakes authors make in this level are as follows:

- The reviewers' comments/feedback are not entirely addressed.
- The authors do not fully comply with the journal's guidelines and requirements for revised papers.

If the revised paper requires a second round of revision, the author will be notified. If the paper is revised successfully, it will advance to the final level.

Level 4: The Final Decision

Level 4 of the communication platform is the final decision. In this level, the paper is inspected for the last time by the editor-in-chief. The paper is then assigned a tentative publication date and the author is notified.

Change in Authorship

JCIS: "Changed Authorship: Please be advised that *JCIS* does NOT allow any re-submission or final submission with changed authorship from the 1st submission."

Authors should not change the authorship from the original submission. Occasionally, change of authorship request is (1) in the order of author appearances for the final paper, and (2) adding an additional author to the final revised paper (who was not on the original submission) who had helped with the revision of the paper.

Before submitting a paper for possible publication in a journal, authors should know that the paper's authorship is final and no changes will occur at a later time. In addition, no new author should be added to the final revised paper regardless.

Final Remarks

A submitted paper must be clear and readable, and with a good writing style that is free of grammar, spelling, and punctuation errors. Often, a paper is written by multiple authors. Each author contributes to various sections of the paper. Each author may have a different writing style that may negatively influence the readability of the paper. The collaborating authors must make every effort to have the paper read by all, and in some cases, have it read by an external person who can provide feedback on the readability and the writing style. Finally, proofread your paper a few times before you submit your paper to a journal.

5

Publishing in Technology and Innovation Management Journals

Perspectives from Both Sides of the Fence

JEREMY HALL

Contents

Introduction

This chapter provides an overview of publishing in technology and innovation management (TIM) journals, specifically why scholars should be interested in such outlets, and how they might improve their chances of success. Scholars are often under pressure to publish in narrowly defined journal lists, many of which do not recognize TIM journals. At the same time there has been a significant increase in submissions to TIM journals, resulting in disappointingly high rejection rates. Part of this increase in submissions may be attributed to the importance of the topic, as well as an overall increased need for scholars from a wider range of countries to publish in credible outlets. I first compare TIM journals with those on commonly used quality lists, followed by some obvious and not so obvious reasons why papers are desk-rejected or do not pass the review process, specifically shedding light on the authors', editors', and reviewers' perspectives. I conclude by suggesting that publishing should be regarded as a "discourse community," composed of authors, reviewers, and editors whose goals and objectives, although varying, are consistent and mutually understood.

I begin this chapter with what would almost certainly result in a desk-reject at any reputable journal: an exaggerated anecdote with no claims of reliability. My first tenure track position started at the same time a new dean took over the faculty. Like any new dean, he wanted to demonstrate his leadership, in this case by raising funds by renaming the business school, and generating recognition by competing in international business school rankings, specifically the London *Financial Times* (*FT*) "Global MBA Ranking."* Many of the *FT* ranking criteria are beyond the direct control of the business school, such as the weighted salary (worth 20%) and salary increase (also 20%); that is, we can't force industry to hire our graduates or dictate their salaries. The dean had more control over the faculty criteria—especially for pretenured professors—such as the percentage of female faculty (2%), international faculty (4%), faculty with doctorates (5%), and the "*FT* research rank" (10%), the number of articles published by the school's current full-time faculty in the now famous (perhaps infamous) "*FT*45" top 45 management journals.

Unfortunately as a male I failed to contribute toward the female faculty criterion, but was able to contribute with a doctorate and as an international faculty member (I have a British passport, albeit a Canadian accent). As a pretenure professor, I was given terse advice to publish in *FT*45 journals, none of which would be considered core TIM journals. To paraphrase my associate dean research, "You can publish as much as you want in TIM journals, but if you go up for tenure and have no *FT*s you will almost certainly be declined." At the time I was an associate editor responsible for environmental innovation for a relatively new sustainable development journal, and was unequivocally informed that I should withdraw: "Don't waste time that could be better spent on publishing in *FT* journals." The *FT* competition evolved into our core mission, "to be a top 50 business school," literally carved in stone on the new granite plaque recognizing the school's new namesake.

---

* The full details of how the *FT* ranks business schools are beyond the scope of this chapter, but are available at: http://www.ft.com/intl/cms/s/2/5728ac98-7c7f-11e3-b514-00144feabdc0.html#axzz3AxPFLRpQ).

The outcome was disappointing. First, as would be predicted in the innovation literature, the *FT* ranking is a moving target. Our first attempt resulted below the target of the top 50, but a respectable score in the 70s, and the dean subsequently left for bigger and better things. Increased participation from other universities in the *FT* rankings resulted in a drop of 10 places the next year, followed by a failure to make the list thereafter, even though our raw scores increased yearly, a true "red queen" effect (Kauffman, 1995). Major pressure was placed on the pretenure professors, where we either changed our employer or our research stripes. I tried the former but did the latter by dropping the associate editorship and focusing on *FT* publications at the expense of TIM journals. I also had bad dreams about being asked to undergo gender reassignment, the remaining criterion I had not fulfilled.

The key themes from this anecdote, and the focus for the rest of this chapter, are as follows. First, there are pressures that discourage publishing in TIM journals, some of which are not in the best interest of the author. Regardless of bad advice from associate deans, TIM journals are legitimate research outlets, and should not be marginalized at the expense of publishing exclusively in journals listed by a ranking scheme such as the *FT*45 list.

Second, journal quality is a moving target where, for example, editorial policies, impact factors, and ranking lists continuously evolve. Indeed, the *FT* list was originally composed of 40 journals; a few have since been added and a few dropped, such as the *Journal of Small Business Management, Long Range Planning*, and *Management International Review*. After I withdrew from the above-mentioned sustainable development journal, it became indexed on the Science Citation Index and in 2014 had a higher impact factor than 32 of the *FT* journals. Under such circumstances a portfolio approach to managing such risk may be better than betting everything on *FT*s; that is, an innovation scholar should try to publish in TIM and other premier publications.

The chapter then discusses the other side of the coin: how to get published when journals receive far more submissions than they are able to publish. Although pressures to publish in elite journals remain, submission rates for TIM journals appear to be increasing at an alarming rate, resulting in disappointingly high rejection rates. I conclude by discussing the obvious, and not so obvious, reasons why papers are desk-rejected or fail to meet the reviewers' expectations, and how this can be reduced by an understanding of the roles and incentives of authors, reviewers, editors, and publishers.

Overview of Technology and Innovation Management (TIM) Journals

This section provides an overview of some TIM journals and how they rank against elite journals recognized by various ranking schemes, with the intent of providing some justification that can be used to convince associate deans and tenure committees why they should be taken seriously.

There are a plethora of journal quality indicators. This includes relatively comprehensive lists compiled from national associations, for example, the German Academic Association for Business Research,* the Australian Business Deans Council,† and the University of Vienna WU-Journal-Rating.‡ Others are only composed of elite journals such as the *FT*45 discussed above, the competing business school ranking *Bloomberg Businessweek* Top 20 (*BW*20) academic journals, and the University of Texas 24 (UT24) leading journal list. Methodologies for determining quality vary (and in some cases are rather vague), resulting in variances in scores for any given journal. Specific details are beyond the scope of this chapter,§ but for now most are based, in varying degrees, on the journal's reputation and impact, which have been measured by, for example, Elsevier's *Scopus* and the *Journal Citation Reports*® (*JCR*) Science and Social Sciences Editions (Thomson Reuters, 2014). The *JCR* provides impact factors (the number of times the average paper has been cited within the prior two years) and other journal quality indicators such as total citations during this period, citations over the last five years, as well as more nuanced indicators.

Most editors and authors are aware of the limitations of such rankings and using indicators as a proxy for journal quality, details of

---

* See http://vhbonline.org/en/about-vhb/. Accessed August 20, 2014.
† See http://www.abdc.edu.au. Accessed August 20, 2014.
‡ http://bach.wu.ac.at/bachapp/cgi-bin/fides/fides.aspx?journal=true;rating=2009). Accessed August 20, 2014.
§ See Professor Anne-Wil Harzing's comprehensive review at http://www.harzing.com/jql.htm

which I discuss below. However, as an editor I realize that a high score is important to many academics. Indeed, I have noticed that such indicators, particularly impact factors, are now used extensively as measures of research quality in, for example, tenure, promotion, recruiting, and grant applications.

The 2013 impact factor of 2.106 for my journal, the *Journal of Engineering and Technology Management* (*JET-M*), more than doubled over last year's score of 0.967, and 2011's impact factor of 1.032. I speculate the improved impact factor can be attributed to a number of reasons. In addition to "statistical" factors (i.e., a few 2012 papers were heavily cited), we transitioned from a manual submission approach to Elsevier's Editorial System (EES), a web-based software for Elsevier journals to manage the editorial process. It is part of Elsevier's business model, and although I occasionally hear complaints from authors and reviewers about it being cumbersome, it provides, for example, automated reminders, allows for more efficient monitoring, and can flag, for example, conflicts of interest in the review process. EES also improves the publication lag, making the articles more current. As can be expected, there is a considerable lag between when a paper is accepted and then cited, which can be expedited through more effective online submission systems. Accepted papers are posted online and citable after about a month of copyediting. Papers thus do not wait in limbo for the next issue, which may be extensive if there is a backlog of papers. We have also been more judicious in desk-rejecting papers that are not aligned with the *Journal*, in part to manage the increasing number of submissions received (discussed below).

Table 5.1 is a sample of TIM journals and their 2013 impact factors listed on the *JCR*. It is based on my interpretation of relevant TIM journals and not intended to be a comprehensive list, but rather for illustrative purposes only. *JET-M* currently ranks third, up from eighth place in both 2012 and 2011. Before I overstate its impact (and authors rush to submit), it should be noted that impact factors often fluctuate over the years, and as a result there is no guarantee that any given journal will retain its ranking. Furthermore, the five-year impact factor, although improved, brings it down to sixth place on the TIM list, whereas citations to other *JET-M* articles are a bit above average. Total citations (648) would pull the journal down to the

Table 5.1   Impact Factors for Technology and Innovation
Management Journals

Technovation	2.704
Research Policy	2.598
Journal of Engineering and Technology Management	2.106
Technological Forecasting and Social Change	1.959
Journal of Product Innovation Management	1.379
Industrial and Corporate Change	1.330
Journal of Technology Transfer	1.305
R&D Management	1.266
Industry and Innovation	1.116
IEEE Transactions on Engineering Management	0.938
Technology Analysis and Strategic Management	0.841
Research Technology Management	0.745
Creativity and Innovation Management	0.714
International Journal of Technology Management	0.492
Innovation: Management, Policy & Practice	0.439
Asian Journal of Technological Innovation	0.167

bottom of the list; as a relatively small journal, *JET-M*'s total citations are significantly less than, for example, *Research Policy* (9,518 total citations), *Technological Forecasting and Social Change* (2,966 citations), *Journal of Product Innovation Management* (2,886), and *Technovation* (2,663). I thus caution authors to avoid becoming too preoccupied with proxies, especially impact factors, which often undergo considerable swings from year to year. As always, the main reason for submitting to a specific journal is because it suits your research and you are able to contribute toward its discourse, which in turn advances its societal impact.

Table 5.2 lists the journals from the *FT*45, University of Texas, and *Bloomberg Businessweek* rankings, along with the top innovation journals (boldface) from Table 5.2 that have an impact factor at least as high as the lowest ranked *FT* journal. As noted above, many readers may have experienced the pressures to publish on these lists, especially for those trying to justify tenure and promotion. The resulting list includes seven TIM journals, none of which are on the elite lists. Two TIM journals, *Technovation* and *Research Policy*, would rank within the top half of the *FT* list and about the middle for the other

Table 5.2 Impact Factors of TIM and Journals Ranked by the *Financial Times* (*FT*45), University of Texas (UT24), and *Bloomberg Businessweek* (*BW*20)

JOURNAL	*JCR* IF	*FT*45	UT24	*BW*20
1. *Academy of Management Review*	7.817	Y	Y	Y
2. *Journal of Finance*	6.033	Y	Y	Y
3. *Quarterly Journal of Economics*	5.966	Y		
4. *MIS Quarterly*	5.405	Y	Y	
5. *Academy of Management Journal*	4.974	Y	Y	Y
6. *Journal of Operations Management*	4.478	Y	Y	
7. *Journal of Applied Psychology*	4.367	Y		
8. *Journal of Marketing*	3.819	Y	Y	Y
9. *Organization Science*	3.807	Y	Y	
10. *Journal of Financial Economics*	3.769	Y	Y	Y
11. *Journal of Political Economy*	3.617	Y		
12. *Journal of International Business Studies*	3.594	Y	Y	
13. *Review of Financial Studies*	3.532	Y	Y	Y
14. *Econometrica*	3.504	Y		
15. *American Economic Review*	3.305	Y		Y
16. *Journal of Management Studies*	3.277	Y		
17. *Journal of Business Venturing*	3.265	Y		
18. *Strategic Management Journal*	2.993	Y	Y	Y
19. *Organizational Behaviour and Human Decision Processes*	2.897	Y		
20. *Journal of Accounting and Economics*	2.833	Y	Y	
21. *Academy of Management Perspectives*	2.826	Y		
22. *Journal of Consumer Research*	2.783	Y	Y	Y
23. **Technovation**	**2.704**			
24. *Journal of Marketing Research*	2.660	Y	Y	Y
25. **Research Policy**	**2.598**			
26. *Management Science*	2.524	Y	Y	Y
27. *Organization Studies*	2.504	Y		
28. *Journal of Accounting Research*	2.449	Y	Y	Y
29. *Entrepreneurship Theory and Practice*	2.447	Y		
30. *Administrative Science Quarterly*	2.394	Y	Y	Y
31. *Information Systems Research*	2.322	Y	Y	Y
32. *Accounting Review*	2.234	Y	Y	Y
33. *Journal of the American Statistical Association*	2.114	Y		
34. *Marketing Science*	2.208	Y	Y	Y
35. *Accounting, Organisations and Society*	2.109	Y		
36. **Journal of Engineering and Technology Management**	**2.106**			
37. **Technological Forecasting & Social Change**	**1.959**			

continued

Table 5.2 (continued)   Impact Factors of TIM and Journals Ranked by the *Financial Times* (*FT*45), University of Texas (UT24), and *Bloomberg Businessweek* (*BW*20)

JOURNAL	*JCR* IF	*FT*45	UT24	*BW*20
38. *California Management Review*	1.944	Y		
39. *Journal of Financial and Quantitative Analysis*	1.877	Y		
40. *Harvard Business Review*	1.831	Y		Y
41. *MIT Sloan Management Review*	1.803	Y		
42. *Production and Operations Management*	1.759	Y	Y	Y
43. *Journal of Consumer Psychology*	1.708	Y		
44. *Journal of Business Ethics*	1.552	Y		Y
45. *Contemporary Accounting Research*	1.533	Y		
46. *Operations Research*	1.500	Y	Y	Y
47. *M&SOM-Manufacturing & Service Operations Management*	1.450		Y	
48. *Human Resource Management*	1.395	Y		
49. **Journal of Product Innovation Management**	**1.379**			
50. **Industrial and Corporate Change**	**1.330**			
51. **Journal of Technology Transfer**	**1.305**			
52. **R&D Management**	**1.266**			
53. *Journal on Computing*	1.120		Y	
54. *Rand Journal of Economics*	1.219	Y		
55. *Review of Accounting Studies*	1.167	Y		
AVERAGE	***2.751***	***2.980***	***3.269***	***3.1229***

two, whereas *JET-M* would rank a respectable 34th on the *FT*45[*] and above four of the journals on the other rankings.

Based on this data, it might be appropriate to advocate for the inclusion of at least one TIM journal on these elite lists. Of course this is a rather superficial comparison; other studies such as Linton and Thongpapanl (2004) and Linton (2009) conducted a more detailed analysis that provides insights on how TIM journals can be assessed. In the meantime I hope that the table can provide incentives for those interested in innovation to publish in TIM journals, and to provide ammunition for their tenure and promotion applications.

---

[*] I once explored why no TIM journals were on the *FT* list and was told informally that *FT* journals were selected based on subject area (e.g., economics, MIS, finance, marketing, strategy, human resource management, etc.) plus practitioner journals, but TIM was not recognized as an area. The closest is perhaps entrepreneurship, with *Journal of Business Venturing* and *Entrepreneurship: Theory and Practice* on the *FT* list but not the UT24 or *BW*20.

How to Improve Your Chance of Publication

The previous section is primarily focused on the author's perspective, and specifically how one might pragmatically justify publishing in TIM journals for career progression. In this section I deal with the other side of the coin: how to get published when journals receive far more submissions than they are able to publish.

In 2011 *JET-M* received 220 submissions, rising to 266 in 2012 and 277 in 2013. As of November 28, 2014, we had 298 new submissions. On average we publish about 25 articles per year. I suspect this high rejection rate is common throughout the field (the manuscript numbers on my submissions to other journals are usually disturbingly high). Part of this increase may be attributed to the importance of the topic, where innovation is often regarded as a panacea for business and societal improvement, a buzzword for corporations, governments, and even business schools and universities. It may also be due to the increased publishing requirements for scholars coming from a wider range of countries. For example, we have had submissions from authors located in over 50 countries.

Many *JET-M* submissions are technical or engineering studies rather than ones suitable for a TIM journal. Perhaps this is due to the title, although potential authors should make the effort to at least scan the journal to determine if their paper is within its scope. Indeed, as an editor the main driver is whether the manuscript engages with, and has an impact on, the *Journal's* discourse. We use a few quick heuristics to help determine whether the paper will be sent out for review. For example, we scan the references, and if there are no citations to TIM journals, then the paper is probably not aligned with our discourse and thus will likely have a greater impact elsewhere. This sounds rather obvious but is surprisingly common, ranging from highly technical papers that have misconstrued our title to manuscripts that appear to have been desk-rejected at mainstream (often *FT*45) management journals. I am sympathetic to these latter submissions, especially if they have been rejected after difficult rounds of revisions. I also appreciate the temptation to resubmit elsewhere quickly and understand the logic of assuming that this will be fine for less-prestigious journals. However, to do so without aligning it with the new journal will almost certainly result in a desk-reject.

Resubmitted papers rejected elsewhere are often flagged by reviewers, who are key players in the publishing process. Indeed, perhaps the biggest editorial challenge is finding good reviewers who provide constructive and timely feedback, and increased submissions have exacerbated this challenge. As a result we desk-reject papers that we believe will not meet reviewer requirements. This includes papers that have obvious methodological flaws, limited discussion of the findings or a dated literature review, as well as papers that do not meet our English language requirements.

Although I believe many scholars sympathize with the difficulties of publishing in a second language, most reviewers do not have the time to struggle through substandard English, and thus usually decline invitations to review. Incoherent titles and poorly structured abstracts are a particular problem, as these are the only sections of a submission included in the reviewer invitation. Inconsistency in formatting, punctuation, and particularly referencing is also important; although most reviewers can appreciate second language challenges, they are not tolerant of sloppiness. More generally a bad experience will result in the reviewer declining future invitations, a scenario that most editors will try to avoid. Unlike authors, editors, and publishers, reviewers have much less incentive to review papers, and as a result reviewing is not a high priority. It is therefore common for reviews to be late, and on rare occasions a scholar will agree to review but disregard the reminders, leaving the paper in limbo. This is particularly a problem if the paper received major revisions, and the reviewer does not respond to requests to assess the revisions. Most reviewers provide excellent feedback, and in most cases authors who fail to respond adequately to reviewer concerns have their papers rejected. Currently at *JET-M* it is much easier to find authors than reviewers, so it only makes sense that reviewers' opinions take priority.

Occasionally a few reviewers recommend citing their own work. Usually this is with justification (that is why they were identified as potential reviewers). However, sometimes it is gratuitous, where the reviewer may be trying to increase their citations superficially in exchange for providing a service to the *Journal* (henceforth "gratuitous reviewer citations"). Note that Thomson Reuters has strict policies

Table 5.3 Top Five TIM Journal Impact Factors: With and without Self-Cites

JOURNAL	WITH SELF-CITES	WITHOUT SELF-CITES
Research Policy	2.598	2.220
Technovation	2.704	2.000
Technological Forecasting and Social Change	1.959	1.483
Journal of Engineering and Technology Management	2.106	1.404
Journal of Product Innovation Management	1.379	1.109

against coercive citing, and journals caught engaging in such behavior can be suspended from the *JCR*. According to Thomson Reuters (2014), "The Journal Impact Factor provides an important measure of a journal's contribution to scholarly communication, and its distortion by an excessive concentration of citations is a serious matter." Self-citing is when editors coerce authors into citing their journal to inflate citations. The *JCR* thus includes an indicator for journal citations without self-cites. Table 5.3 lists the top five TIM journals with and without self-citations. As shown, *Research Policy* and *Technological Forecasting and Social Change* jump up one position each, but otherwise there is relatively little movement. The more opaque "citation stacking," is when, for example, editors from different journals coerce authors into citing each other's journals, and there are now sophisticated algorithms that can detect such behavior.[*] I am not sure if algorithms are currently available that can determine gratuitous reviewer citations, but if not I imagine that it is only a matter of time.

The publishing environment is thus full of sometimes powerful but often indirect incentives and vested interests, and thus ripe for abuse. Although Thomson Reuters and other journal quality indicator companies want to ensure their measures are fair and unbiased, publishers want high-impact journals as a signal of business success. Editors want better quality submissions and the prestige that comes along with an influential journal. Authors want the career benefits of publishing in premier outlets. Indeed, in addition to (quite substantial) tenure and promotion benefits and the flexibility to move to higher-paying positions elsewhere, some institutions provide explicit financial incentives for authors to publish in high-impact journals (Davies, 2011). My original employer discussed in the opening anecdote provided a

---

[*] See http://admin-apps.webofknowledge.com/JCR/static_html/notices/notices.htm. Accessed August 2, 2014.

$5,000 research honorarium for each *FT*45 publication. Deans want the prestige of leading high-quality researchers and scoring high on business school ranking schemes, as well as more functional benefits such as meeting accreditation and external review requirements, and justifying resources from their universities.

Paradoxically, reviewers have the least to gain but have substantial influence over what gets published. Their contributions are usually regarded as a service to the academy and typically do not result in significant contributions toward their annual merit increases. It is thus common for reviewers to regard it as a low priority, or encourage citations to their own work. Conversely authors do not like being dictated as to what they have to cite, especially if it involves changing the paper after what is often a long and sometimes difficult series of revisions. Ironically, reviewers are often drawn from the same pool of authors, yet sometimes seem to forget about their experiences as an author when acting as a reviewer.

Conclusions

Although there are pressures for innovation scholars to submit their best work to (somewhat superficially determined) "elite" journals, doing so may result in a failure to establish a coherent contribution to their specific discourse. Drawing on the *JCR* data, I have tried to provide justification for publishing in TIM journals, which I hope scholars can use to justify their publication strategies to tenure and promotion committees. Note, however, that such proxies have major limitations. Journal quality is a moving target, where editorial policies, impact factors, and in some cases ranking lists evolve. Furthermore, they are only proxies for journal quality, not individual contributions. Citations for individual papers are a more direct measure, although they often take time to accumulate, and as a result may not be a fair measure for junior scholars. (I suspect that a wealth of techniques are now being used to inflate individual paper citations, an area for future research.) Thus, in the long term the safe bet is to submit your research to journals where your scholarly discourse is being set and aligns closely with the readership's interest. More important, a myopic focus on journal proxies rather than the wider intent of the research may very well miss the entire purpose of being an academic.

On the other side of the coin, increasing submissions to TIM journals have resulted in disappointingly high rejection rates. In addition to the specific advice outlined above, a greater awareness of the varying incentives, pressures, and restrictions of those involved in the publishing game can greatly enhance success. A starting point is to become a good citizen in your research community by becoming a reviewer. Although time-consuming, reviewing is perhaps one of the best ways to become familiar with a journal and keep abreast of recent contributions to the discourse, understand current journal trends and interests, and learn how to respond successfully to reviewer comments. It is also much appreciated by the editorial team. Indeed, the temptation for an editor to desk-reject submissions from authors who persistently decline reviewer invitations or provide poor-quality assessments is obviously strong.

Many of my colleagues have tried in vain to publish exclusively in *FT* journals, resulting in a failure to receive tenure or relegation to administration. Others have opportunistically published in *FT* outlets that made their deans happy, but resulted in a confused résumé. For example, many of my former colleagues have published in the *Journal of Business Ethics,* at the time considered to be a relatively easy *FT* publication, even though most could scarcely claim they had any genuine background or interest in business ethics, sustainable development, or other related areas of discourse important to this journal. Others similarly targeted *Management International Review* and *Journal of Small Business Management,* both of which have since been delisted from the *FT.* My point is that although such publishing strategies are understandable for those seeking tenure, in the long term one risks losing one's research identity: it's OK to change your stripes, but don't lose your soul.

Publishing is a tough game, but it doesn't help to get angry, for example, arguing with reviewers rather than attempting to address their concerns, a battle that authors usually won't win. Nor is it helpful to get even, for example, providing unnecessarily harsh reviews or refusing to review for a journal after a paper is rejected. It can, however, be a source of inspiration. For example, the above frustrations I encountered in the opening anecdote shaped my understanding of university research outcomes and resulted in a *Research Policy* publication (Langford et al., 2006) that included the cautionary subtitle "when proxies become goals," a major theme of my research and inspiration for this chapter.

Acknowledgments

Many thanks to Elsevier and *JET-M*'s publishers, Vicki Wetherell, who appointed me as editor-in-chief of *JET-M*, and Jessica Bibb for her continued support, and my dean, Danny Shapiro, for his support and encouragement. Also many thanks to *JET-M*'s managing editor, Vernon Bachor, as well as Stelvia Matos, Bruno Silvertre, and Mike Martin (in memoriam) for their useful comments and insights.

References

Davies, P. (2011). "Paying for impact: Does the Chinese model make sense?" *The Scholarly Kitchen*, available at: http://scholarlykitchen.sspnet. org/2011/04/07/paying-for-impact-does-the-chinese-model-make-sense/ Accessed August 2, 2014.

Kauffman, S. (1995). Escaping the red queen effect. *McKinsey Quarterly*, 1: 118–129.

Langford, C., Hall, J., Josty, P., Matos, S., and Jacobson, A. (2006). Indicators and outcomes of Canadian university research: Proxies becoming goals? *Research Policy*, 35(3): 1586–1595.

Linton, J. (2009). Technology innovation management's growing influence and impact. *Technovation*, 29: 643–644.

Linton, J.D. and Thongpapanl, N. (2004). Ranking of technology and innovation management journals. *Journal of Product Innovation Management*, 21(2): 123–139.

Thomson Reuters. (2014). *The Journal Citation Reports*® Science and Social Sciences 2013 Editions, Thomson Reuters.

6

How Could My Paper Have Gotten Rejected?

JAMES R. MARSDEN

Contents

Introduction

The fact is that many more research journal submissions get rejected than get accepted. Yet another fact is that, as academics, we must succeed in getting our research papers published and recognized (cited) by peer researchers. To succeed, I suggest that one must be able to identify and understand what constitutes those seemingly gray boundary lines between failure and success. I argue that understanding why papers are rejected can actually inform us on why papers get accepted.

Much of the material that follows is not scientific. Although the numerical values used are objective, my interpretation is just that: a subjective view of the numbers, review content, and trends. Being a bit "long in the tooth," I have had quite a bit of experience in evaluating research work and in advising or making decisions on the worthiness of work for publication. In what follows, my focus is on common and continuing issues that I have seen as key elements resulting in so many dreaded negative letters from myself and other editors.

Fit

Perhaps the most likely research paper to be rejected is one submitted to a journal for which the paper is not a good fit. When I can very easily identify a paper as a poor fit, I must admit that I am puzzled. The review process in such a case is easy for me—a simple desk-reject for lack of fit—but I continue to ponder why an author would send us such a paper? As do most journals, we prominently post aims and scope, including a listing of general topic areas. Furthermore, past issues and individual papers are widely available for authors to peruse. Yet, for whatever reasons, papers are submitted that are clearly in domains ranging from mechanical engineering to organizational behavior. The journal that I serve as editor-in-chief (EIC; *Decision Support Systems*) has a focus that is clearly on decision systems or decision support systems and enhanced decision making. We have traditionally welcomed decision-system–focused research drawing from any number of underlying disciplines. This does not, however, imply that papers from the core of such disciplines are within the scope of *DSS*.

The selection of an appropriate outlet can be critical to the acceptance or rejection of your paper. In reading submission letters, I look for evidence that the author has carefully thought through the selection issue and has solid reasons why *DSS* was chosen as an outlet. Surprisingly, some authors fail to mention why they chose *DSS* as a potential outlet or why the journal is a good fit for their work.

As a brief aside, I note that more than a few times I have received submission letters addressed to a different editor-in-chief at a different journal. This suggests to me that the author didn't put in a great deal of time pondering the appropriateness of *DSS* before shipping us his or her latest rejection!

Originality

Since taking on the role of EIC, I have been troubled by the level of professional misconduct by members of our profession. Early in my term, a submitted paper turned out to be one that had already been published in another journal! Even worse, the paper was submitted as a solo-authored piece by one author of the previously published multiauthored paper! The two papers were identical. I have seen an accepted paper in our journal (accepted before my term as EIC but in forthcoming status) withdrawn by an author, only to find out that the paper had been accepted and published in another journal.

Such occurrences are, thankfully, rather rare. But issues of plagiarism and repetition of earlier published work are not so rare. My view is quite clear and rests upon two key factors. First, the journal is not interested in publishing previously published work; we are interested in new and original work. Second, even when the earlier work is that of the author him- or herself, this does not overcome potential copyright issues. When we sign copyright transfer forms, we no longer own the copyright. Submitting a paper that repeats material directly from our own prior work is not OK because, in most cases, we had transferred the copyright to the publisher. An author may seek permission to repeat a figure or a diagram (with proper release forms and clear annotation to the repetition). As an EIC, when I find direct repetition of earlier work, the potential copyright issues lead me to reject the submission quickly with an explanation of the problem. In the following paragraphs, I expand on two particular recurring problems in this arena of replication.

In literature reviews, it is not uncommon to see direct lifting of summary paragraphs from earlier work. Rather than explaining and synthesizing prior related research, some "authors" apparently find that effort too taxing and, instead, provide a set of carefully lifted summary text directly from the work of others. Quotations are, of course, not used. After rejecting a paper that exemplified this problem, the author sent me an indignant e-mail noting that the earlier paper was listed in the references. Sometimes, the words of previous authors are incredible in their clarity and strength of reasoning. In such cases, repeating a short passage verbatim might prove valuable to the readers. That is fine, but that is what quotes are for! These words are the work of others. We need to credit those authors and their work clearly.

A second common problem involves the extension of a previously published conference proceedings. Many, if not most, conference proceedings are copyrighted and authors sign a copyright transfer form. However, even for the cases where there is no copyright transfer for a proceeding, the fact is that the material has already been published. As noted earlier, as an EIC, I have no interest in republishing earlier published work. Upon receiving rejections for such papers, I have had authors lecture me about "standard practice" and even a "30% new material standard," neither of which is consistent with any copyright law with which I am familiar. In our field, conference proceedings tend to get minimal, if any weight, in promotion or tenure reviews. Furthermore, given that a variety of working paper outlet options are available, why do we even need conference proceedings? The current expansion of web posting services is another option for seeking input and encouraging discussion. One issue I do have with the postings is the length of time a paper is posted. At what point is the work no longer fresh, no longer "new"?

Screening tools are becoming more sophisticated and can search amazing arrays of prior work. I use (or, for accuracy, I should say my editorial assistant uses) these tools on every submission. I long for the day when this is not necessary, but I doubt that such a day will come. To date, I have not seen a noticeable decline in the rate of submissions having red flags raised by the screening process.

Writing

From my experience, there is a clear, unwritten yet pervasive, heuristic operating throughout the reviewing process: a well-written paper gets the benefit of the doubt! For a poorly written paper, the opposite holds. These are outcomes that I personally support. Poor writing and poor presentation give the impression that the author just didn't care enough about the work to put in the time and effort to present the work well. Poorly written manuscripts can be difficult to wade through and understand. Isn't it natural that reviewers (and editors) tend to get frustrated and perplexed at why the author didn't exert the energy to deliver a quality manuscript?

Without claiming that the situation is "fair," the current fact is that major journals in IT (information technology) and IS (information systems) are predominantly in English. This puts an extra burden on the many academic researchers for whom English is a second (or third) language. I continue to be impressed by the writing quality of the best work from non-English–speaking countries. This shows an extra level of effort and care, elements that pay off in clarity of the work and the likelihood of success. On the flip side, I continue to be shocked that bad writing (and poor spelling!) comes frequently from native English speakers. In many cases, there is little evidence that the author(s) took any time to edit or even proofread the manuscript carefully before submission. The review team is not there to perform the functions of professional copy editors. The review team is there to assess the value and contribution of the research. When a paper is difficult to wade through and to understand, even a high-value actual contribution is likely to go undiscovered.

Quality writing is the responsibility of authors and not reviewers. Few, if any, of us are great writers. We all must continually strive to improve. Reviewers and editors are there to help, to make suggestions for enhancing the flow and clarity of the presentation. They are not there to write or rewrite a paper for a sloppy author.

But what does good writing entail? Most importantly, I would stress brevity and clarity. Be brief and be clear in your presentation. It is crucial that very early in a paper you clearly delineate the research question you address. You must also carefully and clearly explain the contribution of your work. Professional editing help is available at many universities. On many occasions, either as an EIC or just in conversation with colleagues, I have heard the woeful refrain that the "reviewers just didn't understand my work!" My retort (often unspoken to avoid reflexive verbal turmoil) tends to be, "Whose fault is that, the author's or the reviewer's?" Even with great writing, confusion may occur, but I would believe that the outcome pairing good writing and confusion is quite infrequent compared to the pairing of poor writing and confusion.

I highly recommend that authors take advantage and get their work professionally edited. Seeing and discussing recommended edits or changes can be an important part of the continuing effort to improve writing skills. The researcher must help the reviewers through clarity in writing and presentation.

Definition of Terms and Constructs: Measurement

Although the definition of terms and concepts along with processes of measurement may certainly be viewed as part of good writing, I decided to key it as a separate section in this presentation. I do this for two reasons: the critical importance of this content, and the surprising frequency of authors ignoring this importance. If a theory or a proposition is to be subjected to test or initial validation, then each variable or construct utilized must be clearly and fully defined. There must be clear operational definitions so that subsequent researchers can repeat or extend the testing and validation processes.

I have been troubled by how often precise operational definitions of constructs are not provided in a manuscript. Constructs are discussed and linked to prior usage in the literature, yet are often not defined. The situation is not helped by the shifting of definition or interpretation or measurement to third parties such as survey respondents. Recently, I read through an assortment of papers that asked respondents to answer questions regarding whether their organization had "strong IT planning capabilities," "skilled IT staff," and "the knowledge necessary for deploying IT applications." Yet no operational definitions or measurement processes were provided to those sent the survey (at least no evidence was offered by the author that these had been provided). So what is respondent A "measuring"? What is respondent B measuring? What do any responses actually measure? Without concrete definitions and clear measurement processes, the "data" gathered is no more than subjective interpretations.

Need for Validation

One set of almost surely doomed submissions involves little more than what I term "professional conjecturing." These papers come in many forms, but most leave reviewers with "readers' remorse" and result in negative recommendations. A key element that cuts across the variety of such papers is the lack of careful validation. Let me explain using a few (actually common) examples:

- *Papers delineating, and only delineating, a framework:* I do not suggest that frameworks cannot be helpful, but they must be shown to be useful. It is not enough to speculate that a framework might help. There must be clear validation through rigorous investigation and evidence that the proposed framework actually delivers enhanced decision making (or, in some settings, useful and valuable information). As a brief aside, I have yet to actually see a definition in any of these papers of the critical term, "framework."

- *Papers that present a pure mathematical model utilizing heroic assumptions:* Again, I am not arguing that mathematical modeling is not valuable. Rather, I am suggesting that developing a heroically simplified model actually is very likely to remove the possibility of validation. Can the authors of such papers actually identify a market where "all consumers are identical" or a market where "there is one seller, one product, and n consumers uniformly distributed across a preference line" or a market where "the single seller knows the preferences of the n consumers"?

Speculation and conjecture can certainly play important roles in research idea generation. Interactive research discussions comparing and contrasting research ideas, approaches, methods, and methodologies are often important elements in initiating and developing research. They are elements in the research process, but they are not by themselves the output. There must be validation that clearly demonstrates the applicability and usefulness. Speculation and conjecture can initiate investigations that deliver new and significant contributions. But those contributions must be demonstrated and not merely alleged or assumed.

There is a set of important papers that focus on development of new techniques, new processes, and perhaps new algorithms. Here, too, validation is important, although a bit different from what was discussed just above. In these latter situations, the author must demonstrate why the proposed techniques/processes/algorithms are better than current counterparts. What is gained? What is the cost to achieve these gains? Presenting a new technique is not sufficient. A broad analysis of the improvement or superiority of the new process must be provided. Providing an example or two where the new process or technique works or yields some gains provides little of value. One can easily construct convenient (rigged?) examples. The author should seek to identify the set or range of problems or conditions where the new process outperforms existing counterparts.

Contribution

So suppose (OK, I have revealed myself as a trained economist!) we receive a submission whose subject matter appears to be a good fit for the journal. Further assume (again, the economist reveals himself) the manuscript is all original, very well written, and contains careful validation. Is the author home free? Not yet because a crucial factor remains, a factor that is the true decisive element. The real issue is, "Does the paper make a significant new contribution to the field?"

We work hard to identify reviewers who are experts in the research domain of the submitted paper, reviewers who are in strong positions to understand and identify a new contribution and determine its significance. Here again, however, the author has the responsibility to help the reviewers "see the light." Helpful techniques include a carefully structured and well-synthesized literature review that delineates what has been done and what has not been done. By the end of the literature review section, the author should have led the reader to ponder, "I wonder why no one has investigated 'X'? 'X' is clearly an important issue with significant implications." The author then guides the reader through an understanding of how the presented research analyzes "X" and what important findings will be demonstrated in the paper.

The requirement to demonstrate a new and significant contribution results in the death knell for a large set of formulaic papers. In the past few years, one set of such papers has involved the concept of intention to use one of any number of emerging technologies.

I did a quick Google Scholar search of "intention to adopt" "technology." Limiting the results to those only from 2013, the search yielded 1,670 results (as with all Google Scholar searches, this list includes many unpublished manuscripts). The standard approach tends to include a broad survey (sometimes web-based, sometimes not) along with a very standard array of processes (composite reliability analysis, Cronbach's alpha analysis, convergent and discriminant validity analysis, common method bias analysis, and finally structural modeling). To be successful, the author of such a formulaic paper must demonstrate to the review team that there is a clear, new, and significant contribution in performing the analysis on online banking services versus online shopping services, or between online banking services in Taipei versus online banking services in Barcelona. Judging from recent review outcomes and viewpoints expressed by reviewers and AEs (associate editors), these are not easy hurdles. In many cases, the only difference between some papers is how the survey was administered.

Certainly subjectivity is present when reviewers or editors consider whether a manuscript presents a significant new contribution. One reason for multiple reviews and editorial analyses of each paper is to seek a balanced overall analysis. In the end, however, it is the author's responsibility to explain clearly the specific contribution and its value.

And, Finally, a Few Things to Avoid

After taking on the role of EIC, I must admit that I was rather dismayed by various communications I received from submitting authors. The vast majority have been very professional. But there have been several that do make me wonder whether the sender understands (or even has tried to understand) the academic journal review process. Here are a few examples of communications that are examples of what not to e-mail:

"I am anxious to hear—when will I receive the reviews?"—*This was received seven days after submission.*

"I need to know by next week"—*This was received three weeks after submission.*

"The reviews are unfair—obviously, new reviewers were used for the revision. Reviewer 1 who performed the two rounds of reviews is not the same person since the reviewer has strangely shown a lack of understanding"—*In fact, Reviewer 1 was the same Reviewer 1 as in earlier rounds and had tired of the author failing to respond to the issues raised and suggestions offered. In fact, I believe that Reviewer 1 displayed excellent understanding.*

"The reviewers say that they did not find a significant contribution. What is the significant contribution that is not evident in the manuscript??"—*Anyone want to answer this one?*

"The rejection is certainly not something we can accept"—*I don't think any commentary is necessary on this one.*

I could go on but these examples should be sufficient. Most inquiries I receive are very professional. Sometimes the process does take longer than I would like and a status inquiry is quite reasonable. Rejections no doubt hurt (I know my own rejections have hurt!), but such outcomes should not lead to lashing out. Valid points and issues can easily get buried by a nasty tone.

As authors, we need to understand and accept the reality of the publication process in academic journals. I offer the following brief list of "things to remember" as an author involved in the process:

- *Reviewers, AEs, SEs (senior editors), and most EICs are unpaid volunteers who generously give their time to support the advancement of research in our fields.* I am continually pleasantly surprised at how willing academic researchers are to give freely of their time. Quality reviewing requires significant time and effort that takes reviewers away from their other activities. The generosity demonstrated by so many is a testament to how seriously these individuals take the concept of professional responsibility.
- *Reviewers (and even editors) are just as busy as authors.* We can't expect a reviewer to drop everything when our paper arrives in his or her e-mail box; reviews take time; we strive for a 90-day average turnaround from submission to decision, but turnaround times do have variation running from much less than 90 days to somewhat more than 90 days. Our colleagues—the reviewers—do travel, do get sick, and do have many demands on their time.
- *Reviewers can (and quite often do) disagree.* Even with our best efforts, the reviewing process retains subjectivity. One reviewer may see a flaw that another does not. One reviewer may see an important implication or possible use of the research that another does not. We utilize multiple reviewers (at least two and often three), an SE, and an EIC to seek fairness and weed out individual bias. Yet differences certainly occur. In the end, whether the decision is to accept or to reject, we must remember that these are, in fact, professional judgments. These judgments are neither right nor wrong, but rather our best professional judgments.
- *The issue of fit for a particular journal is a dynamic rather than static issue.* A key early issue facing an EIC or SE is whether a submission is a good fit for the journal. Over time, the focus of the journal and the interests of the readership do change. The relevance and importance of topic areas do change over time. An author has the responsibility to delineate why the submitted paper is a good fit for the journal at the present time.

- *Each editorial decision is no more than our best professional judgment.* Can EICs, SEs, AEs, and reviewers make bad decisions? Certainly they can and do in both directions, that is, accepting papers that should be rejected and rejecting papers that should be accepted. As EIC, all I can ask from the review team are best professional judgments. I may not personally agree with the outcome, but if I don't respect the efforts and recommendations, how many members of the team will continue to find participation in the review process a worthwhile use of their time?

Some Final Thoughts—Things We Can Do Better

In the end, the quality of a journal and its review processes reflect all of those involved with the journal: the authors, the reviewers, and the editorial team. As well as we might think we perform each of our roles, we all can do better. In the preceding discussion I have focused on the authors and detailing how the probability of success can be increased. I would be remiss if I didn't outline suggestions on the reviewer side.

It is important to remember that authors and reviewers come from the same pool of individuals. We always hope, and indeed expect, that successful authors will agree to be reviewers. In recruiting peer reviewers, I like to stress three important characteristics that reviewers should exhibit:

Fairness: Review the work as you would like your work to be reviewed.

Open-mindedness: Each of us has greater or lesser familiarity in differing methods and methodologies; the important issue is not whether I would use the same methodology as the submitting author(s), but rather are the methods or methodologies appropriate to address the research question and are they correctly applied?

Constructive criticism: A reviewer serves as a professional critic, but the criticism should be constructive. Comments and suggestions should emphasize and explain ways in which the work can be improved. Although a reviewer may recommend rejection, the goal of the reviewer should be to assist the author(s) in improving the work.

There is an implicit element of fairness that I have been very pleased to see exhibited by many invited reviewers, that is, the importance of avoiding conflicts of interest or the appearance of conflicts of interest. This is no easy task because our community, although perhaps seeming large, is actually quite small. The invited reviewer is in the best position to declare a potential conflict of interest. EICs, SEs, and AEs work to avoid invited reviewers with clear conflicts of interest; however, we are limited in our knowledge. Each time I receive an e-mail from an invited reviewer declaring a potential conflict of interest, I very much appreciate this action. In addition, I gain added respect for these individuals and keep them in mind for future editorial team openings.

Finally, if you are not an active reviewer, I urge you to become one. Reviewing takes a lot of time but offers a true learning experience. It is also a critical part of our professional responsibility. In the end, quality reviewing is the heart of academic research publishing. Decisions must be made on what work excels and is ready for publication and what work is not. The process is not perfect but I must admit it works well most of the time.

In a recent editorial in the journal,[1] I urged authors to be accepting of decisions. I stressed that, although I know I (and all of us) make mistakes, final decisions have to be made. With apologies "to the *Rubáiyát of Omar Khayyám* and substituting EIC for 'moving finger'," I offered the following which I repeat here:

> [T]he EIC writes; and, having writ, Moves on: nor all thy Piety nor Wit Shall lure it back to cancel half a Line, Nor all thy Tears wash out a Word of it.[2]

Endnotes

1. J.R. Marsden, "Transition II," *Decision Support Systems*, 64: 1–3.
2. "The Rubaiyat of Omar Khayyam," Omar Khayyam, January 1, 2009, Cosimo; online, http://www.library.cornell.edu/colldev/mideast/okhym.htm

7

ADVICE ON PREPARING AND REVISING JOURNAL MANUSCRIPTS IN BUSINESS AND SOCIETY TOPICS

DUANE WINDSOR

Contents

Introduction

Publishing scholarly research in academic journals involves a set of specific skills, which can be studied continuously and honed with experience. The set of skills, beyond value-added scholarly content which is necessary but not sufficient, includes manuscript preparation, manuscript revision, deconstruction and developmental reviewing of other authors' work, and neutral professionalism in communications with action editors, reviewers, and other authors. The chapter uses the term "action editor" to refer to any of the possibilities of editor in chief, coeditor, or associate editor for journal decision. The author draws on experiences as editor, author, and reviewer, together with some available information on journal process and developmental reviewing, to offer advice on this set of skills.

Business and society as a field of study covers a broad range of research topics. Journals in the field publish work on business ethics, business–government relations, corporate governance, corporate social responsibility and performance, environmental management and policy issues, stakeholder theory, the international and cross-national comparison dimensions of those topics, and related topics. Journals therefore receive manuscripts from specialists trained in a diverse array of disciplines including (but not restricted to) anthropology, economics, finance, law, marketing, organizational studies, philosophy, political science, psychology, and sociology. This diversity is both an opportunity for authors and a problem for editorial judgment and reviewer selection.

This chapter provides advice to authors on preparing and revising journal manuscripts and explains the key reasons why manuscripts are rejected at the first stage of assessment—initially by an action editor—or later by reviewers, and how to reduce the likelihood of such rejection. The author explains how he handles revision requirements. He also explains a number of aspects of journal process including editorial judgment concerning reviews, time for reviewing and revising, relationship between electronic and paper publication, and care in avoiding plagiarism or appearance of plagiarism errors. The chapter also provides advice on how to prepare developmental reviews of other authors' manuscripts. Learning to write well and to review well are closely related skills in academic publishing.

The writer edited the peer-reviewed academic journal *Business & Society* (founded 1960) for eight years (2007–2014). He has published in *Business & Society* (*BAS*), *Business Ethics Quarterly* (*BEQ*), and the *Journal of Business Ethics* (*JBE*), among other journals, as well as provided reviews for the *Academy of Management Journal*, *Academy of Management Review*, *Strategic Management Journal*, and others. The author also draws on comments by others made at panels during the period 2007–2014 of editors and authors without direct attribution, as the author generally concurs with such comments as are paraphrased here and such comments may have been expressed over the years by more than one person. The International Association for Business and Society (IABS), the academic association sponsor of *Business & Society*, and the Social Issues in Management (SIM) and the Organizations and the Natural Environment (ONE) divisions of the Academy of Management sponsor such panels at their annual conferences for doctoral students and junior faculty as an aid in helping young scholars to develop their understanding of and skills in academic publishing.

In recent years, IABS and SIM have also been sponsoring manuscript development workshops at their annual conferences for providing more direct advice to young scholars from more senior scholars. These manuscript development opportunities can be very valuable to young scholars.

Business & Society, founded in 1960 and in Volume 53 in 2014, is sponsored by IABS, which appoints the editor. The journal is owned and published by Sage Publications, which is an independent international publisher of journals, books, and electronic media for scholarly, educational, and professional markets. Sage is the world's fifth largest publisher of journals, publishing something more than 700. *Business & Society* entered the Thomson Reuters Social Sciences Citation Index® in December 2009, and received its first five-year impact factor (*Journal Citation Reports*) in July 2014. During 2007 to 2013, *Business & Society* published quarterly with a total budgeted length of 696 printed pages. In 2014, the journal increased to six issues (bimonthly) with an increase in total length to 870 pages. Both regular articles and special issues or forums are published through this fixed cycle. (Additional special issues are not published.) As of July 2014, the journal shifted to a system of coeditors and a managing editor to increase speed in handling of the increasing volume of submissions (both regular and special issue) by spreading the workload. On January 1, 2015, the new coeditor team assumed full editorial control.

Manuscript Preparation and Submission

Personally, this editor thinks that the value-added content of a manuscript is the overriding consideration. A manuscript should make some useful contribution to theory (such as addressing a significant gap in the extant literature) or add to the body of empirical knowledge in a topic area (whether confirming or disconfirming present information). Ideally, the contribution or addition should be more than simply marginal, even if typically incremental rather than revolutionary. On this general view, value-added content in a poorly prepared and written manuscript in principle can be developed toward publication through reviewing and revision under editorial guidance; but a well-prepared and written manuscript of little content merit is not worth developing toward publication. Publishing is both a skill in scholarship (value-added content) and a skill in writing (a well-prepared and readable manuscript).

A basic consideration in submitting to a journal is to bear in mind that journals, and editors and reviewers, vary considerably. Generally, the stronger the journal (in terms of academic quality, academic reputation, and metrics such as impact factor), the lower the acceptance rate. In 2005, a decade ago, *Academy of Management Journal* reported about a 30% desk-reject ratio, as typical of top-tier organizational and management journals, and about an 8% natural (rather than targeted) acceptance ratio for both regular and special research forum submissions (From the Editors, 2005, p. 734). The *AMJ* editors then indicated that they preferred to increase the number of articles published rather than reduce the acceptance ratio due to the increased volume of submissions (From the Editors, 2005, p. 734). For the 2011 cohort of regular submissions to *BAS*, the author's hand-calculated information is about 50% desk-reject ratio and a 15% acceptance ratio rounded up slightly. At *BAS*, the desk-rejection assessment has been handled through the editor or a coeditor, an associate editor, or a lead reviewer consulted by the editor. The *BAS* shift to six issues with more pages reflected both increasing volume of submissions and a lengthening queue of articles waiting for a paper issue slot. The desk-reject rate may tend to increase and the acceptance rate to decrease with these conditions at a journal.

The necessity for desk-rejection is straightforward. Editor and reviewer effort are valuable resources to be focused on manuscripts most likely to be accepted for publication. Authors naturally want acceptance, or at least developmental input that improves the odds of being accepted at the next journal. Journals are not the proper forum for developing papers, although considerable developmental support is in fact provided in working toward which reviewable manuscripts should be accepted or rejected.

A first requirement for submission of a manuscript in highly competitive conditions is to prepare the manuscript itself properly according to the format and style preferences of the journal. These preferences vary greatly across journals. An author should look at the stated preferences and also a recent article from the target journal. Journals are generally pretty clear about format and style. Not being consistent with these preferences can be an immediate reason for desk-reject, or an instruction back that the submission will not be assessed further until the stated preferences are met. Journals may vary in how strictly they enforce format and style preferences up front. So this first requirement is not automatically a minimum requirement everywhere. However, an author submitting to an academic journal should recognize that the odds are strongly stacked against acceptance, and therefore it is simply good practice to prepare the manuscript as properly as possible so as to safeguard better against upfront enforcement. As a reviewer, the present author typically looks at the citations and references to see if they are reasonably uniform and then reasonably consistent with the journal's expectations. Nonuniform citations and references signal that the submitter is casual in manuscript preparation. Casualness begins to trigger for a reviewer, or editor, significant reservations about how serious the submitter is, and also how well the submitter is likely to be able to handle revision work if the manuscript is to proceed. First impressions are important and can last for reviewers and editors. No submitter is perfect, especially on the initial draft submitted; the standard is not perfection but reasonable evidence of seriousness.

An author should make the initial draft of a manuscript for journal submission as sound as possible in terms of content. There are limits to a single author's capacity to do so; the real purpose of an external, double-blind reviewer is less how to reject a manuscript and more how to help develop a manuscript (From the Editors, 2005, p. 734). At strong journals, manuscripts are likely to go through multiple rounds of review and revision. The stronger the manuscript at the onset of this process, the better its chance of being selected for revision and for getting through the multiple-round process to acceptance and publication. However, content is no guarantee of acceptance at top-tier journals. At *AMJ*, 2002 information (cited in From the Editors, 2005, p. 732) suggested that low level of contribution (as judged by the review process) was one of the top two reasons for rejection of nearly five-sixths of the submissions. Given this reality—and the BAS editor has similar experience—creating a bad first impression with a poorly prepared manuscript is not the way to proceed.

A good method for studying manuscript preparation, which this editor has heard expressed by more than one panelist, is to deconstruct a published article from the target journal or by a well-regarded scholar. The purpose of this deconstruction is to take the article apart to see what makes it succeed. How did the author prepare the article in terms of presentation as well as content, and especially the abstract, the introduction, and the ending material? The first part of a paper is about the hook: what makes the paper interesting and novel. One can and should study the success of others.

Desk-Rejection and Rejection after Review

Because the odds for rejection are so high, the important first step is to survive through desk-rejection or rejection after review to a revision invitation. This section explains the key reasons why manuscripts are rejected at the first stage of assessment, by an action editor or later by reviewers, and how to reduce the likelihood of such rejection.

There are several bases for desk-rejection by an action editor, depending on the system used by the particular journal. Inadequate manuscript preparation, lack of reasonable fit to the journal's scope, and underdevelopment are the most typical bases. Desk-rejection manuscripts are generally not worth much detailed assessment effort and feedback, because the problems are fundamental.

Inadequate manuscript preparation, in terms of format and style, is one basis, as discussed in the previous section. At *BAS*, sometimes if the content seems strong and the problem is manuscript format and style (especially in citation and reference materials), the action editor may elect rarely to return the manuscript for corrective work. (A corrected manuscript can be substituted into the online submission system before reviews are commissioned.) Otherwise, if the content seems weaker, the submission may simply be rejected on the grounds that the developmental process for the manuscript appears problematic. Top-tier journals are likely not to waste their effort: the manuscript should be ready for assessment.

Two other bases for desk-rejection concern fit to the journal and underdevelopment of the manuscript's content. For *BAS*, which has a quite broad scope of interest, fit is a highly subjective decision by the editor. The journal has certain core interests, listed earlier. The more distant a submission is from those core interests, the less likely the journal is to decide to review that submission. A manuscript that is really a finance paper or a marketing paper (and such submissions do sometimes arrive), with little connection to *BAS* core interests, is typically desk-rejected with advice to go to a more appropriate journal. A highly philosophical paper may be more suitable for *Business Ethics Quarterly* or *Philosophy of Management*.

One consideration in fit is the reference set. A manuscript should be an addition to a conversation within the journal. If the reference set does not contain much in the way of prior articles published in the target journal, then questions arise concerning whether (1) the manuscript fits the journal, or (2) if there is fit, whether the author has bothered to consult the journal's prior publications. For *BAS*, *BEQ* and *JBE* may be close substitutes in this regard. For instance, a marketing-oriented manuscript that cites largely marketing literature may thus prove problematic for these nonmarketing journals. There is an increasing literature on marketing and corporate social responsibility, and thus *BAS* might have interest in joining that research stream. However, the decision ultimately rests on the value-added contribution to the *BAS* scholarly community rather than on the value contribution to the marketing literature (for which there are journals).

An underdeveloped manuscript is one that sketches an idea or reports, fairly briefly, data findings but typically lacks a strong theory section, literature review, and implications section. *BAS* more commonly receives manuscripts (double spaced) of about 40 pages length. Manuscripts of say 20 pages length, that seem not to be suitable even as research notes (which might run say 25 pages), are very likely to be underdeveloped. A manuscript can also be much too long, as papers much more than 50 pages at the outside tend to prove unwieldy. Both short and long manuscripts may be returned with a decision of "reject but redevelop and resubmit," meaning that the paper needs to be developed or shortened for further assessment. Typically an action editor will provide some limited guidance for an original manuscript that has not been reviewed; and more guidance if the decision is reached with external reviews.

At *BAS*, given roughly a 50% desk-rejection ratio and a 15% acceptance ratio (for regular submissions), then about one-third of the regular manuscripts are reviewed and rejected either at the first or second rounds of reviewing. (The second round is a review of a first revision.) *BAS* tries, in principle, to reject no later than the first revision review process in order not to delay authors unnecessarily. Top-tier journals may reject at even a third round (review of a second revision), and possibly even subsequently. The review process, ideally, should be developmental in helping to improve a manuscript materially, even if ultimately the decision is to reject. Action editors are trying to make informed judgments about value-added contributions, and for manuscripts that should be reviewed they cannot make such judgments without advice from external reviewers. The review panel, comprising an action editor and a minimum of two external reviewers (and quite typically three reviewers), is ideally trying to formulate as much information as possible about the strengths, weaknesses, and prospects of a manuscript within a topic area.

This assessment process is uneven and not very predictive of future impact on scholarship. "The inescapable conclusion is that authors, editors, and reviewers are poor judges of predicting high-impact papers at the time the paper is accepted for publication in a peer-reviewed social science journal" (Lewin, 2014, p. 169).

Handling the Manuscript Revision Process

When an author receives a "revise and resubmit" invitation or a "reject but redevelop and resubmit" invitation, then the revision effort is the critical step. Given the low likelihood of acceptance, the first revision is typically a make-or-break point. Revision is a learnable skill that is different from initial manuscript preparation skill. There are two levels of "revise and resubmit": reasonably standard and "high risk." Standard revision, although not a guarantee of success, tends to signal that the editor and reviewers are as a panel in general agreement about the likely value of a manuscript and how to revise it for the next round of reviewing. High-risk revision tends to signal that one or more reviewers, and perhaps the editor, express serious reservations about the likely value or likelihood of successful revision. High risk signals typically no better than 50–50 likelihood of success.

A first step is to deconstruct the editor's guidance (the decision letter) and the reviewers' comments (see Liu, 2014). Ideally, the editor provides some reasonably detailed advice on how to undertake revision. On rare occasion, this editor has found that the reviews are sufficiently clear and consistent to recommend that the author follow the reviews in preparing a revision. But generally, editors try to give detailed instructions and also address any obvious conflicts or other difficulties in the reviews. In deconstruction, the author should break the editor guidance and the reviews up into numbered points (as in Editor, #1, #2, #3, ...; Reviewer A, #1, #2, #3, ...). Then the author can see which points overlap or are in conflict; and order the points in importance or criticality for revision. Generally, authors will want to work from more critical to more minor, on the basis that the minor points are useless to handle if the more critical points cannot be handled. Conflicting requirements from reviewers should be referred back to the editor for guidance, if this matter was not spotted and addressed earlier in the editorial guidance.

After deconstructing the editorial guidance and reviews, the present author prepares an author response document (commonly required by journals) and organizes a table with the numbered item and details on the handling in the manuscript. Typically, this table will be organized as Editor, Reviewer A, and so forth. The present author then works back and forth between the revision in process and the author response document. The response document may be almost as important as the revised manuscript in persuading the editor and reviewers of the merit of the revision. On occasion, an author may encounter a requirement or recommendation with which the author disagrees. In general, an author should try to satisfy the editor and reviewers. But when the author is convinced that the requirement or recommendation should not be followed, then the appropriate procedure is to document in the response document the author's view on the matter. Editors and reviewers may take into account the author's reasoning. Otherwise, an action editor should insist on a particular course of action, explaining why and sometimes how.

Action editors are, in principle, attempting to reach a balanced judgment about a manuscript, including during the revision process. This process requires open-minded reviewers on the one hand and assertive (but not undiplomatic) authors (Tsang, 2014). The action editor wants to make the published article as good as possible. The author should strive for the same goal. A good action editor and professional reviewers are partners with authors in this developmental process. Revision work is more complicated with teams of coauthors. There is a recent report that range of expertise and good communication among coauthors may improve the likelihood of acceptance (Rupp et al., 2014).

The present author has, as an editor and a reviewer, seen a large proportion of author response documents that he would describe as unnecessarily unctuous. The author thanks the action editor and each reviewer for support, for favorable comments, and for useful suggestions. This approach has the appearance of being highly concerned to cultivate the action editor and reviewers in favor of the manuscript. The present author thinks that the review and revision process should be a professional, diplomatic, and open (if double-blind) conversation about how best to prepare a particular manuscript for publication. On rare occasion, the present author has thanked an action editor or reviewer for a particular point, because it was particularly insightful and useful and simply had not occurred to this author. Otherwise, the author response document should be to the point.

Two issues that have received increasing attention concern frivolous self-citation and coercive citation. Some authors appear to follow a pattern of citing their own previous work in manuscripts frivolously, meaning to no useful point (Lynch, 2010). Generally, at *BAS*, the editor has removed self-citations when obviously revealing of authorship before sending a manuscript out for external review; self-citations tend to breach the double-blind standard. For an accepted manuscript, sometimes self-citation is useful and often it is not. A judgment is involved: is the self-citation really useful; if not, it should not be used. On a related point, citation impact factors tend to include self-citation with a journal to other articles published in the journal, but reported information also includes proportion of self-citations. Much above 30% self-citation is problematic for a journal: much lower is better.

Coercive citation means the action editor or reviewer in effect requires or strongly recommends particular citations not necessarily directly relevant to strengthening the manuscript (Wilhite and Fong, 2012). Sometimes reviewers recommend their own work for consideration. Sometimes editors may want to increase the self-citation count of the journal (although as noted above, there is some risk in an increased self-citation ratio). A proposed safeguard (Wilhite and Fong, 2012) is that a journal self-citation should always be combined with a citation from another journal. The present author has never encountered this problem as an author. A difficulty in assessing this problem is that a manuscript may not have effectively included the prior relevant literature in a journal. As pointed out earlier, one consideration in deciding whether to desk-reject a submission is whether there has been any conversation on the subject matter in the article and that judgment is reached by looking at the references.

Additional Matters

A good way to learn how to write better is to review better. A good and professional review should be developmental (Hempel, 2014), rather than simply critical. The present author's approach as a reviewer is generally to try to identify the strengths and the weaknesses of the manuscript (however good or bad the manuscript on the whole), and then to offer concrete recommendations for how to address the weaknesses so as to improve the manuscript. The present author also tries to identify as many minor matters as possible as a help, such as citation–reference system, typos, grammatical errors, and so forth. Although this approach requires a close reading of a manuscript and attention to detail, the approach also hopefully assists the action editor in making a decision and helps the author in future revision work (whether for the target journal or elsewhere). This approach to reviewing can help a junior scholar learn how to deconstruct a manuscript for the purpose of self-improvement in writing. A single author does not have the advantages possible in team coauthorship.

It is important to take care to avoid plagiarism errors. Scholars know that plagiarism is prohibited (https://www.informs.org/Find-Research-Publications/INFORMS-Journals/Author-Portal/Publications-Policies/Guidelines-for-Copyright-Plagiarism). This editor has encountered two kinds of plagiarism errors other than the obvious misconduct of copying whole paragraphs from another source without attribution. One error may arise in sloppy citation documentation practices, at least as much as or perhaps more than in deliberate taking of material from other sources without attribution. In preparing or revising a manuscript, an author working in some haste takes information or material from another source but forgets to be sure that the citation information is included. A good rule of thumb is that a string of six words in a sequence should be in quote marks and cited explicitly. In paraphrasing rather than quoting, generally the source should be cited. Some care should be exercised in this regard. Once a reviewer identifies a likely situation of missing citation information (with the possible allegation of plagiarism, intentional or inadvertent), the action editor has little choice but to ask the author to withdraw the manuscript. A second error lies in an arguably grayer area of using verbatim suggestions from reviewers. The reviewers are familiar with their information and are likely to see such usage on the next round of reviewing. Reviewers may reasonably assert that the language was contributed and thus deserving of attribution in some way or at least paraphrasing rather than using without modification. A good practice is to rewrite reviewers' contributions in the author's own language, and to add some acknowledgment in a specific footnote or in the general acknowledgment statement about use of reviewers' contributions.

Generally the best practice is to inform the editor of the target journal of what is involved with a submission. Duplicate submission—to two or more journals simultaneously without notice to the editors—is prohibited by *BAS* policy (and commonly so among journals). Editors, and reviewers, sometimes discover undisclosed duplicate submissions. Redundant publication, whether essentially the same article or reuse of the same data without much value-added contribution, is often undesirable, and journals will generally opt against publishing such work when the editor is not informed of the circumstances. (There can be limited circumstances under which journals may elect to publish what is partly, or even largely, the same information.) Another practice is sometimes termed "salami slicing." The term conveys that a research project is sliced up into multiple papers for submission to different journals. An absolute general rule is difficult to write, but the basic principle is that slicing is not desirable. Again, however, the best practice is to inform the editor at the outset of the submission (by cover letter or e-mail). Reviewers do uncover such slicing, in the form of having reviewed an earlier part or version or having read an earlier version in another journal (online or in print).

Journal process varies considerably across journals, depending on the resources available to the particular journal. Generally, top-tier journals shoot for 50 to 60 days average time to a first decision. However, this metric as an average is highly misleading. Desk-rejects may occur within a week; and getting a manuscript fully reviewed and a decision letter out may take three months or even much longer in practice. *BAS* has gone from a single editor with a part-time assistant to a new system of coeditors and a managing editor precisely in order to speed up. Each manuscript proceeds on a different schedule, depending particularly on availability of qualified reviewers. Top-tier journals can enforce reviewer deadlines by effectively delisting late or weak reviewers from future invitations. Other journals are not in this fortunate position, but must attract reviewers and encourage them to work according to a similar schedule without much incentive beyond the desire to help develop a field of study.

Many journals, including *BAS*, now publish articles electronically (online with a DOI) in advance of paper publication. The idea is to get articles circulating as quickly as possible. Authors should undertake more effort to notify colleagues and scholarly communities of the availability of articles. Generally, they can post abstracts to listservs and websites. The present author's view is that acceptance is what matters, and the general quality of the journal issuing the acceptance, rather than when a paper issue appears or the resulting citation count. There may be universities that require page numbers for an article (meaning a paper issue, as electronic online versions for paper journals generally do not include unique page numbers in a numbered issue). However, the present author's view is that such requirements are archaic.

The final point is to encourage all authors to treat action editors and reviewers with neutral professional courtesy and diplomacy. Junior scholars in particular are under a lot of pressure to publish for reasons of institutional tenure and promotion decisions. Dissatisfaction with slow and erratic processes is quite understandable, but should be expressed diplomatically and with an understanding for context. This editor has always encouraged authors to contact *BAS* by e-mail about every three months to track status. Journals are constrained by resources, mostly manpower availability. Action editors and reviewers are just people, carrying out a responsibility that is almost universally not compensated financially. These volunteers are also teaching, conducting research, and serving on committees. An action editor can get a desk-reject decision out quickly. Developmental effort toward acceptance for publication can take a couple of years or more depending on how quickly the action editor and reviewers can work, and how quickly and effectively authors can revise. Then there can be a significant lag for copyediting and production, electronic publication, and paper publication. An author should have a pipeline of works so that disruptions or delays in one journal can be handled within the author's career development schedule.

References

From the Editors. (2005). Everything you've always wanted to know about *AMJ* (but may have been afraid to ask). *Academy of Management Journal,* 48(5): 732–737.

Hempel, P.S. (2014). The developmental reviewer. *Management and Organization Review*, 10(2): 175–181.

Lewin, A.Y. (2014). The peer-review process: The good, the bad, the ugly, and the extraordinary. *Management and Organization Review*, 10(2): 167–173.

Liu, L.A. (2014). Addressing reviewer comments as an integrative negotiation. *Management and Organization Review*, 10(2): 183–190.

Lynch, J.G. (2010, December 13). "Frivolous Journal Self-Citation," http://ama-academics.communityzero.com/elmar?go=2371115

Lynch, J.G. (2012). Business journals combat coercive citation. *Science*, 335(6073): 1169.

Rupp, D.E., Thornton, M.A., Rogelberg, S.G., Olien, J.L., and Berka, G. (2014). The characteristics of quality scholarly submissions: Considerations of author team composition and decision making. *Journal of Management*, 40(6): 1501–1510.

Tsang, E.W.K. (2014). Ensuring manuscript quality and preserving authorial voice: The balancing act of editors. *Management and Organization Review*, 10(2): 191–197.

Wilhite, A.W. and Fong, E.A. (2012). Coercive citation in academic publishing. *Science*, 335(6068): 542–543.

8

INTERDISCIPLINARY RESEARCH

Pathway to Meaningful Publications

DONALD E. BROWN

Contents

Introduction

Interdisciplinary research provides the opportunity for teams to address complex problems from multiple perspectives. The importance of research that crosses the boundaries of traditional disciplines is now widely recognized by funding agencies and universities. Often a vice president for research, dean, or department chair will set out to show that their university, school, or program reaches beyond traditional academic boundaries to mine the interstices between fields and disciplines. Just because you hear this talk of interdisciplinary activity, the questions always linger. Can you get grants doing interdisciplinary work? Can you get promoted doing it? Can you even publish in it? It turns out the answer is yes to all these questions, but successful interdisciplinary research and publication requires employing a thoughtful approach to integrating components of the different disciplines. This chapter gives you guidance for effective integration strategies that can produce significant research results and publication success.

For much of history, scholars studied problems not disciplines. This meant that scholarship consisted of broad knowledge of natural phenomena, language, abstract reasoning, and more. In the seventeenth century Edmond Halley was not only a first-rate observational astronomer but also the developer of a diving bell that he himself tested in the river Thames. We remember Gottfried Wilhelm Leibniz for his development of the infinitesimal calculus with notation that we continue to use today. However, Leibniz also wrote extensively in philosophy, politics, and law, among other areas, and his contributions to these fields are still recognized.

Suffice it to say that modern scholarship demands considerably more specialization than it did in the seventeenth century and earlier. The academic disciplines that we know today are products of only the last 150 years and, as such, represent a relatively new way of organizing humanity's quest for understanding. The compartmentalization of scholarship into disciplines makes it difficult for single individuals to acquire both the depth and breadth of knowledge needed to tackle the most pressing problems. Instead teams of researchers from multiple disciplines now conduct the groundbreaking work on important questions.

An unintended consequence of this specialization by discipline is the concern and possibly even bias in academia against interdisciplinary work by faculty members, particularly, junior faculty members. Recent doctoral graduates and new faculty members often receive advice that interdisciplinary research is not well regarded by hiring and tenure and promotion committees. These new faculty members are also told they should only participate in interdisciplinary projects after obtaining tenure (Lattuca, 2001; National Academies, 2014). Pfirman and Martin (2010) identified difficulties publishing interdisciplinary research in traditional disciplinary journals as an obstacle that must be overcome in order to facilitate interdisciplinary research. In a review piece, Jacobs and Frickel (2009) concluded that although it is unclear if interdisciplinary articles face a citation penalty, they are definitely disadvantaged in the sense that the interdisciplinary journals in which this work tends to be published are typically of lower status than disciplinary journals.

Evidence suggests that these concerns are not only misplaced but also potentially counterproductive. In fact, it is narrow disciplinary research that can both limit opportunities and stifle meaningful results. Lattuca

(2001) notes that disciplines "delimit the range of research questions that are asked, the kinds of methods that are used to investigate phenomena, and the types of answers that are considered legitimate." Furthermore, disciplines impede the study of problems of importance to society and to our greater understanding of complex phenomena (Becher, 1989; Kuhn, 1970; National Academies, 2014). As to the impact of interdisciplinary publications on academic success, Jacobs and Frickel (2009) studied a 2008 survey of doctorate recipients and found that a higher number of publications in interdisciplinary research do not have a "dramatic" effect on the types of positions individuals hold. Their study also showed that doctoral interdisciplinary research actually increases the likelihood that the candidate will obtain an academic position.

Both funding agencies and universities have also recognized the importance of interdisciplinary research. The National Science Foundation (NSF) web page states: "NSF gives high priority to promoting interdisciplinary research and supports it through a number of specific solicitations. NSF also encourages researchers to submit unsolicited interdisciplinary proposals for ideas that are in novel or emerging areas extending beyond any particular current NSF program" (National Science Foundation, 2014). Many universities are linking existing programs in departments to create new interdisciplinary offerings as well as starting new organizations with interdisciplinary mandates (Brint, 2005). Clearly, interdisciplinary research has become a valuable and important part of the modern university.

Despite the recognized importance of interdisciplinary research, to conduct work along the boundaries of different fields remains a challenging and frankly scary prospect for even senior faculty members. To provide an overview of how to begin to navigate these challenges, the next section describes key features of disciplines that must be accommodated to make interdisciplinary research and publishing successful. The following section then provides case studies and examples of effective approaches to linking disciplinary activity to produce distinctive research results and publications.

Interdisciplinary Research

The National Academies defines interdisciplinary research as "a mode of research by teams or individuals that integrates information, data, techniques, tools, perspectives, concepts, and/or theories from two or more disciplines or bodies of specialized knowledge to advance fundamental understanding or to solve problems whose solutions are beyond the scope of a single discipline or area of research practice." There are related concepts called multidisciplinary and transdisciplinary, but for our purposes we focus on interdisciplinary and use the National Academies' definition as the framework for this discussion.

The Academies' report identified four drivers for interdisciplinary research: "the inherent complexity of nature and society, the desire to explore problems and questions that are not confined to a single discipline, the need to solve societal problems, and the power of new technologies" (National Academies, 2014). More recently several transformational events have made interdisciplinary work critical to scholarship in science, technology, and industry: (1) the continued unabated acquisition of unprecedented amounts and types of data; (2) the development and deployment of all the components of an integrated, scalable, and sustainable cyber infrastructure; and (3) the emergence of powerful mathematical and statistical modeling frameworks for the combination and integration of evidence from disparate sources and types. These three transformational events have already altered the way we conduct research in science, engineering, and business and they are beginning to change the way we teach in these fields.

To meet the challenges represented by these transformational events researchers have actively begun to pursue a wide range of activities with the goal of making progress against many of the important problems that reside at the borders between disciplines. The National Academies' definition of interdisciplinary research puts the entirety of this border-crossing process into a single verb: *integrates*. But integrating knowledge from different disciplines creates enormous challenges. To understand these integration challenges it is instructive to explore what we are integrating, namely the essential components brought by each discipline.

Lattuca (2001) notes that defining a discipline can take two forms: "They can be defined as sets of problems, methods, and research practices or as bodies of knowledge that are unified by any of these. They can also be defined as social networks of individuals interested in related problems or ideas."

The first definition is structural and the second is cultural. A number of researchers have characterized the structural composition of disciplines (e.g., Dressel and Marcus, 1982). Kuhn's characterization provides both the structural components as well as insights into the surrounding cultural milieu. Kuhn (1970) described the framework or paradigm of disciplines in terms of several components: (1) generalizations or theory, (2) models or synthetic examples, and (3) exemplars. Kuhn's description of the interplay between these components shows how the culture of a discipline evolves and sustains itself.

Of course, the problem when working at the boundaries between disciplines lies in the clash between generalizations, models, and exemplars. Disciplines differ in the importance they place on each of the three components, which in turn provides insights into how accepting they will be of interdisciplinary research and publications. For example, disciplines such as computer science and business place low value on generalizations and theory whereas for other disciplines such as physics and mathematics, theory is the most important component. For still others, for example, economics and engineering, the value of theory is somewhere between the extremes. Any attempt at integrating work between disciplines requires acceptance of mutually agreeable generalizations, models, and exemplars. In fields where the epistemological structures of these components are stable and accepted, more work is required to enter into interdisciplinary challenges. On the other hand, disciplines with less stable structures are more open to engaging with other disciplines and working toward new opportunities to extend the epistemology of their own field and to contribute to that of others. For an extensive discussion of these points see Becher (1989) and Lattuca (2001).

In addition to the structural and cultural barriers between disciplines, perhaps an even more fundamental barrier is communication. Members of a discipline speak to each other in a technical language or jargon that enhances communication among its members but impedes communication between disciplines. As with translation between natural languages, attempts to translate between the technical speech of disciplines are not reducible to simple recoding of symbols or expressions. Early studies have shown that poor communication between interdisciplinary team members is a cause for failure of research projects (Niles, 1975; Kuhn, 1977; Bauer, 1990).

Because the key to publication success is research success, we need approaches to overcome these very real challenges to effective interdisciplinary research. With that understanding, publication success will follow as we show with examples in the next section. The communication challenge is foremost for without improving communication structural alignment among the represented disciplines will be impossible. To enable effective communication the researchers must learn enough of the jargon in the other fields to permit concise but accurate information exchanges. They should also work to employ a common language that can be understood broadly among members of the separate disciplines. The time spent in learning these new languages and doing these translations will not only help ensure research success but also will pay off handsomely when it comes time to write publications. If the researchers effectively communicate with each other they will in turn be able to communicate in their publications to members of their respective communities. They will then have the options of multiple publication outlets:

1. Interdisciplinary journals where the results are communicated in a common language that many fields can understand
2. Multiple disciplinary journals where the results are communicated in the language of the respective disciplines

Once the researchers have developed an understanding of the basic language the disciplines use, they can turn to the challenge of structural mismatches. These mismatches can occur in the generalizations, models, or exemplars contained in the paradigms of the different disciplines. The first step to resolve these mismatches is simply to understand that they may exist. In the rush of working together on an exciting problem, the researchers may fail to realize until they are far into their own part of the problem that their colleagues have different assumptions, accepted examples, or modeling techniques.

Starting with the realization that they have mismatches will allow the researchers to look for points of disagreement as the second step. Actually they should look for areas of commonality as well as mismatches. They are looking for these in the relevant theory, models, and exemplars in each discipline. Finally, they should take their findings and exploit the commonalities while searching for the sources of the differences. In most instances the differences can be resolved by understanding that they proceed from different but mutually acceptable sources. In those rare instances where they are not resolvable that way, there are three possibilities: (1) there is actually no mismatch; it is a communications problem; (2) the sources of the mismatch have not been truly uncovered; or (3) one or both of the disciplines has something wrong. Needless to say, if the third alternative turns out to be the reason for the mismatch, these researchers have just made a major and important discovery (e.g., see the history of plate tectonics; Hughes, 2001).

Taken together these strategies for successful interdisciplinary research imply one thing: hard work. In fact a two-year study of technology assessment projects done by teams from multiple disciplines showed that the greater the diversity within the teams, the more successful the project integration (Rossini et al., 1979). This ran counter to the study's hypothesis and the investigators then concluded that greater disciplinary diversity may have incentivized the teams to work harder to achieve integration.

Interdisciplinary Research and Publication Examples

This section provides examples and case studies of the use of the methods described in the previous section for successful interdisciplinary research and publications. We start with examples of successful communication and then proceed to structural integration.

Interdisciplinary Communications

Examples of good communication between researchers that resulted in interdisciplinary publications can be found in journals that seek and publish interdisciplinary work. The *IEEE Transactions on Systems, Man and Cybernetics: Systems* is one of these journals and a recent publication illustrates the principles of successful communication between disciplines (Combi et al., 2014). The title of the article is "Representing Business Processes Through a Temporal Data-Centric Workflow Modeling Language: An Application to the Management of Clinical Pathways" and the keywords are healthcare systems, information systems, time management, and workflow management. The paper combines results from systems engineering, computer science, project management, and healthcare. Knowledge of the language of healthcare and medicine is evident when the authors write, "a laparoscopic intervention may need the results of the concurrent bioptic analysis to be properly concluded while exceptional recovery activities have to be performed in case of emergency evidence during standard treatment; however, the successful application of a fibrinolytic therapy requires a maximum delay of 30 min after the admission into the emergency department." They also express the problem in technical terms from systems engineering and project management: "Current workflow systems are lacking in effective management of three general key aspects that are common (not only) in the clinical/ health context: data dependencies, exception handling, and temporal constraints." Finally, they refer to computer science concepts when they write, "Moreover, we analyze the computational complexity of the temporal controllability problem in TNest, and we propose a general algorithm to check the controllability."

These authors have learned and melded the technical languages of multiple disciplines to both conduct and describe the outcomes of a highly interdisciplinary project. In this particular paper, all four authors have degrees in computer science, but two of them have postdoctoral appointments in the Department of Public Health and Community Medicine, University of Verona, and have learned the language of that field. This paper provides a good example of successfully integrating the technical languages of multiple disciplines.

Interdisciplinary Structural Matching: Exemplars

According to Kuhn (1970) and others (Lattuca, 2001), disciplines have their own exemplars or specific instances where generalizations and models meet problem-specific data. These exemplars provide validity checks not only on theory and modeling of the discipline but also on understanding by members of the discipline. "One of the fundamental techniques by which the members of a group, whether an entire culture or a specialist sub-community within it, learn to see the same things when confronted with the same stimuli is by being shown examples of situations that their predecessors in the group have already learned to see as like each other and as different from other sorts of situations" (Kuhn, 1970).

When working in interdisciplinary teams shared exemplars provide a vehicle for maintaining a common direction to the work and validate the efficacy of the work to the separate research communities. Essentially shared exemplars provide an efficient means of communicating the effectiveness and importance of the work to the different disciplines involved in the research.

As an example of interdisciplinary research that uses shared exemplars in their publications to reach audiences from multiple disciplines, consider the work of Moorman and his colleagues on monitoring intensive care unit patients to predict adverse events (Moorman et al., 2011a,b). This research group has members from a variety of different disciplines to include medicine (specifically cardiology and pediatrics), statistics, mathematics, and engineering (specifically signal processing). They highlight the importance of their interdisciplinary approach when they write that "teams of clinicians and quantitative scientists can work together to identify clinically important abnormalities of monitoring data, to develop algorithms that match the clinicians' eye in detecting abnormalities, and to undertake the clinical trials to test their impact on outcomes" (Moorman et al., 2011b).

This group has a number of published papers but the two referenced here illustrate the team's use of shared exemplars. One of the papers was published in the proceedings of an engineering conference, the 33rd Annual International Conference of the IEEE Engineering in Medicine and Biology Society, whereas the other paper was published in a medical journal, the *Journal of Pediatrics*. So they are successfully publishing in multiple disciplines, which shows that they have effectively presented their results to the reviewers in the separate disciplines. This means that they are not only communicating well between members of the separate disciplines but more important, they have also matched one or more of the paradigmatic components of the different disciplines.

To illustrate that this matching has occurred with exemplars we quote from their writings: "We developed heart rate characteristics analysis in 4 years, and then spent nearly 7 in a randomized trial. Randomized trials are the only way to convince clinicians to change their practice, and this is how it should be..." (Moorman, et al., 2011b). Clinical trials are the gold standard exemplars for not only clinicians but more generally for medical research. By using this gold standard of medical exemplars their work has been widely published in the medical literature. This team also uses exemplars for statisticians and engineers in the form of frequency domain parameters, phase domain plots, signal quality indices, and entropy estimates. These exemplars have ensured acceptance by the journals in these fields. The combined use of these exemplars from the different domains has helped these researchers cross the boundaries of the different disciplines involved in this work and to publish broadly to a wider audience not possible with single disciplinary research.

Interdisciplinary Structural Matching: Theory

Astronomy and chemistry are two disciplines with well-formed theory and, hence, they are less permeable to interdisciplinary work. To bridge these two fields requires careful understanding of the respective theoretical constructs in each discipline and where bridging opportunities between them exist. Researchers in the new interdisciplinary field of astrochemistry have discovered these opportunities and their work represents a good example of collaborative research that effectively integrates theoretical constructs between disciplines. Example papers in this area are Niles (1975) and Öberg et al. (2011).

Researchers in astrochemistry are motivated to fill the current gap in our understanding of the emergence of chemistry in the universe. One of the greatest achievements of nineteenth and twentieth century chemistry was the development of mechanistic organic chemistry. The systematic quantitative study of the structure and reactivity of organic molecules has provided chemists with the blueprints to construct new complex molecules with important physical, medicinal, and materials properties. Astrochemists build upon this established theoretical foundation to study the initial synthesis of molecules, especially organic molecules, from the elements and from small, abundant interstellar molecules. They have built a two-way bridge from chemistry to astronomy by studying chemical physics and organic chemistry in the interstellar medium.

The unusual and inhospitable conditions of the interstellar medium require nature to be creative in the ways it constructs molecules. A combination of theory and experimental physical chemistry is needed to test the viability of a wide range of novel synthetic routes. In the gas phase, these include ion–molecule reactions, radiative association, and low-temperature tunneling reactions. In addition, cosmic ray and extreme ultraviolet processing of the interstellar ices coating silicate nanoparticles are thought to generate reactive species that can undergo barrierless reactions to build larger molecules. Although these mechanisms may be exotic, they have created nature's largest reservoirs of chemically bonded matter in the space between the stars.

In this new field of astrochemistry an understanding of mechanistic interstellar chemistry becomes essential for astronomers to understand the structure and evolution of astronomical objects. If a common chemical evolution associated with the star and planet formation process can be identified, then the chemical composition—its molecular fingerprint—of the object becomes a new way to identify the stage of astronomical evolution. More broadly, the next-generation radio astronomy interferometers provide a fundamentally new way to observe and understand chemically rich astronomical objects. Instead of using the electromagnetic spectrum emitted by the object, the dominant way of studying objects in the universe, the "chemical image" can be constructed by substituting molecular species for colors of light.

The researchers in this area clearly demonstrate the power of interdisciplinary research to tackle grand challenge problems such as the emergence of chemistry in the universe and the evolution of star formation. However, the integration of disciplines with well-established, formal theories such as chemistry and astronomy is only possible with carefully linked theoretical constructs from each discipline. The hard work that this linkage requires is justified by the promise that this collaboration brings to tackle truly grand challenge problems.

Interdisciplinary Structural Matching: Models

The last of Kuhn's paradigmatic elements that can be matched for interdisciplinary work is modeling. Disciplines use models to connect their generalizations or theories to data. As with theory some disciplines have well-established, formal modeling methods that change only in the presence of new data and, possibly, changes to the theory. Other disciplines are more accepting of modeling methods from a variety of fields.

Papers by Brown and his colleagues illustrate the matching of modeling methods from systems engineering, operations research, law enforcement, and criminology. Their work has integrated predictive models from these fields to show areas threatened by various types of crime (Poter and Brown, 2007; Wang and Brown, 2012).

Predictive models from law enforcement and criminology use database management systems (DBMSs) and geographic information systems (GIS) to show hot spots or areas where lots of crime has occurred. However, in general these models tend to be reactive rather than proactive. A more proactive approach requires early warning of trouble with sufficient lead time to formulate a plan. Early warning, in turn, necessitates the development of predictive models in space and time that can inform law enforcement of pending hot spots rather than historical hot spots.

The work by Brown's team builds predictive models that integrate the DBMS and GIS from law enforcement with other available data such as census and social media. They then build mathematical models that capture the decision making of criminals through functional relationships between demographic, economic, social, victim, and spatial variables and numerous measures of criminal activity. By modeling decision making they can predict criminal activity as the variables in the environment change.

To evaluate these models in a manner acceptable to the multiple disciplines involved, Brown's team took a number of steps to assess the effectiveness of the models and to show their usefulness within the different disciplinary cultures. To illustrate this consider the team's modification of the standard approach to assessing estimation accuracy or correctness in prediction. A commonly used method for model evaluation in engineering and statistics dating to the Second World War is the receiver operating characteristic (ROC) curve. This curve plots true positives against false positives. A true positive is when the model predicts an event and it happens, whereas a false positive is when it predicts an event and it does not happen.

Brown's team modified this classic approach to model evaluation to make it more acceptable to law enforcement and criminology. They created the surveillance plot which shows true positives plotted against the area covered by surveillance. The surveillance area covered is the percentage of the total area that contains a police patrol or other surveillance capability that can detect and suppress the next criminal event. So if the model requires the police to cover 100% of the area in order to predict 20% of the events correctly, then it is not doing very well. On the other hand a model that allows police to cover only 5% of the area and still predict the location of the events with 50% true positives does very well.

This kind of model integration is quite necessary for successful interdisciplinary work among disciplines where theories are not well formed and modeling is used to understand and test theories. Criminology is an example of such a field. The models that Brown's team built allow the criminologists and law enforcement specialists to test and evaluate competing theories of criminal behavior. They also allow systems engineers and operations researchers to evaluate their methods for optimizing predictive accuracy and estimation using real data. So this example gives good insights into how interdisciplinary teams can develop integrated modeling to the benefit of the disciplines and to enable new types of problem solving.

Conclusions

As the report of the National Academies (2014) indicates, interdisciplinary research has "delivered much already and promises more." As a number of studies have shown, interdisciplinary research does not adversely affect academic success or publication record and, in fact, may be the path to truly important and satisfying work. However, this path is not easy. It will require the researchers minimally to invest time in understanding the technical languages of the participating disciplines. For even more substantial contributions the researchers will need to work to link the paradigmatic elements, theory, models, or exemplars from the different disciplines. The choice of linkage will depend on the disciplines involved and the nature of the problems studied. This process is not as easy as staying comfortably within the confines of one's discipline but the results from engaging in interdisciplinary research can be truly significant for the investigators and for society.

References

Bauer, H.H. (1990). Barriers against interdisciplinarity: Implications for studies of science, technology, and society (STS). *Science, Technology, & Human Values*, 15(1): 105–119.

Becher, T. (1989). *Academic Tribes and Territories: Intellectual Enquiry and the Cultures of Disciplines*. Bristol, PA: Society for Research into Higher Education and Open University Press.

Brint, S. (2005). Creating the future: "New directions" in American research universities. *Minerva*, 43: 23–50.

Combi, C., Gambini, M., Migliorini, S., and Posenato, R. (2014). Representing business processes through a temporal data-centric workflow modeling language: An application to the management of clinical pathways. *IEEE Transactions on Systems, Man, and Cybernetics: Systems*, 44(9, Sept.): 1182–1203.

Dressel, P. and Marcus, D. (1982). *Teaching and Learning in College*. San Francisco: Jossey-Bass.

Hughes, P. (2001, February 8). "Alfred Wegener (1880-1930): A Geographic Jigsaw Puzzle." Retrieved August 22, 2014, from *On the Shoulders of Giants*. Earth Observatory: http://earthobservatory.nasa.gov/Features/Wegener/wegener_2.php

Jacobs, J.A. and Frickel, S. (2009). Interdisciplinarity: A critical assessment. *Annual Review of Sociology*, 35: 43–65.

Kuhn, T.S. (1977). *The Essential Tension*. Chicago: University of Chicago Press.

Kuhn, T.S. (1970). *The Structure of Scientific Revolutions*. Chicago: University of Chicago Press.

Lattuca, L.R. (2001). *Creating Interdisciplinarity: Research and Teaching among College and University Faculty*. Nashville, TN: Vanderbilt Univerity Press.

Meland, L. (2010). *The Marketplace of Ideas: Reform and Resistance in the American University*. New York: Norton.

Moorman, J.R., Carlo, W.A., Kattwinkel, J., Schelonka, R.L., Porcelli, P.J., Navarret, C.T., et al. (2011a). Mortality reduction by heart rate characteristic monitoring in very low birth weight neonates: A randomized trial. *Journal of Pediatrics*, 159(6, Dec.): 900–906.

Moorman, J.R., Rusin, C.E., Lee, H., Guin, L.u., Clark, M.T., Delos, J.B., et al. (2011b). Predictive monitoring for early detection of subacute potentially catastrophic illnesses in critical care. In *33rd Annual International Conference of the IEEE EMBS*. Boston: IEEE Press, pp. 5515–5518.

National Academies. (2014). *Facilitating Interdisciplinary Research*. Washington, DC: National Academies Press.

National Science Foundation. (2014). "Introduction to Interdisciplinary Research." Retrieved August 22, 2014, from National Science Foundation: http://www.nsf.gov/od/iia/additional_resources/interdisciplinary_research/

Neill, J.L., Steber, A.L., Muckle, M.T., Zaleski, D.P., Lattanzi, V., Spezzano, S., et al. (2011). Spatial distributions and interstellar reaction processess-patial distributions and interstellar reaction processes. *Journal of Physical Chemistry A,* 115: 6472–6480.

Niles, J.M. (1975). Interdisciplinary research management in the university environment. *SRA, Journal of the Society of Research Administrators,* 6(9): 9–16.

Öberg, K.I., Qi, C., Fogel, J.F., Bergin, E.A., Andrews, S.M., Espaillat, C., et al. (2011). Disk imaging survey of chemistry with SMA (DISCS): II. Southern sky protoplanetary disk data. *Astrophysical Journal,* 734(2): 12.

Pfirman, S. and Martin, P. (2010). Facilitating interdisciplinary scholars. In R. Frodeman, J. T. Klein, and C. Mitcham (Eds.), *The Oxford Handbook of Interdisciplinarity.* Oxford, UK: Oxford University Press, pp. 387–403.

Poter, M. and Brown, D.E. (2007). Detecting local regions of change in high-dimensional criminal or terrorist point processes. *Computational Statistics and Data Analysis,* 51(5): 2753–2768.

Rossini, F.A., Porter, A.L., Kelley, P., and Chubin, D.E. (1979). Frameworks and factors affecting integration within technology assessments. In R. Barth and R. Steck (Eds.), *Interdisciplinary Research Groups: Their Management and Organization.* Seattle: Interstudy, pp. 136–158.

Wang, X. and Brown, D.E. (2012). The spatio-temporal modeling for criminal incidents. *Security Informatics,* Springer. 1:2. http://link.springer.com/article/10.1186%2F2190-8532-1-2#page-1

9

MODELS OF EDITING AND EDITORIAL BOARDS

DANIEL E. O'LEARY

Contents

Introduction

Journals are distinguished from each other by a number of factors that ultimately can influence the size and structure of the editorial board, the "way" that editors edit and other editorially related issues. This chapter investigates some of these issues ultimately generating models of editorial boards and the editorial process.

This chapter is particularly concerned with editorial boards inasmuch as they provide organization and support to editors. Although editorial boards exert influence on their editorial activity, editors probably seldom choose the model of their editorial board.[*] Even when editors are part of a start-up journal the editorial board format and number of members is likely to be copied from some existing journal as a benchmark.

As a result, the purpose of this chapter is to provide some models of editing and editorial boards in order to begin generation of a theory of editing and editorial boards. In particular, this chapter captures key variables from the editing and editorial processes and examines the potential impact of those variables on editing and editorial boards. Specifically, this chapter builds a model where three independent variables (journal size, whether they are supported by an association, and the base discipline) are related to two dependent variables (number and organizational structure of the editorial board). Such a model would allow empirical analysis of a key supporting structure for editors—the editorial board.

---

[*] This and other empirical questions in this chapter need empirical assessment.

Variables for a Model of Size and Structure of Editorial Boards

The purpose of this section is to discuss the independent and dependent variables of a model of journal editorial boards. Such a model could provide the initial basis of a theoretical and empirical study of academic journals, their editorial boards, and other characteristics.

Independent Variables: Size, Association, and Discipline

This section lays out some of the potential independent variables that could influence the editorial variables of size and editorial board structure. In particular, this chapter suggests that those independent variables include the size of the journal, whether the journal is associated with a society or other sponsoring group, and the "discipline" that the journal functions in, for example, computer science or psychology. First, different journals vary by size, where size could be measured by many variables. For example, size could be measured by the number of papers or pages submitted each year or the number of papers or pages published in each year in the journal.* That size variable can affect a number of operational activities associated with editing a journal, because, in general, the more papers there are to process the more people are required to process them. Second, journals are published by different sources, ranging from societies (IEEE or ACM) and special interest groups to independent publishing groups (e.g., Taylor & Francis, Elsevier, etc.). Those societies and special interest groups have members who receive the journal, often as part of being a member in the organization. Accordingly, in some cases, the journals have been established to provide a particular perspective or to provide their members with a "voice" or representation. Third, journals are a part of virtually every academic discipline. Specific disciplines or leading journals in those disciplines could provide editorial models from which journals model themselves. Such existing journals in the discipline are likely to provide models of best practices or ways

---

* Size could be measured from an input perspective (e.g., papers submitted), an output perspective (e.g., pages published), and from the relationship between inputs and outputs (papers submitted/papers published).

of doing business that are at least partially related to the particular discipline. Thus, the discipline can provide models of editorial boards and editing.

Dependent Variables: Editorial Board Size and Structure

There are at least two editorial variables that appear to be dependent on those independent variables. Each of these two variables is readily visible, typically on the inside cover of the journal or on the journal web page, where the editorial board and structure are given. Perhaps the most visible aspect of that editorial board is the number of editors. Analysis of editorial boards suggests that the total number of editors on those journal editorial boards varies substantially. For example, the total size of the editorial board can range from 10 or 20 members to 100 or more.

Editorial boards typically provide a formal structure to support the editor. The organization of editorial boards can range from a single editor and his or her editorial board to multiple layers of editors, such as associate and assistant editors. In addition, editorial boards can include advisory boards or other structures. Journals and their editorial boards provide a peculiar type of organization with particular editorial board structures and sizes. As such, they could provide important empirical data that could be studied in hopes of understanding editorial board best practices or optimal organizational forms.

Editorial board structure influences or reflects the day-to-day operation of the journal. In more hierarchical editorial organizations, the editor often does not actually edit papers. Instead the editor assigns the paper to an associate editor who is responsible for managing the referee process. In some settings, the associate editor makes a publication recommendation, whereas in other settings the associate editor makes the final decision.

Editorial Board Model

This section examines the relationship between the variables provided in the previous section to generate a model of editorial boards that can be investigated empirically. The model is summarized in Figure 9.1.

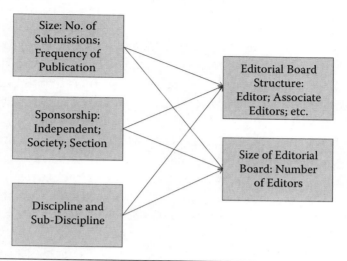

Figure 9.1   Model of editorial boards.

Effect of Size

A priori the size of the journal can be expected to influence the number of editors required to process those papers. For example, if there are a larger number of papers rather than a smaller number of papers there is likely to be a need for more editorial capabilities (more editors). Furthermore, as the number of papers increases, rather than having the editor make every decision, associate editors could be given the authority to make decisions about papers. Accordingly, as the demand to process papers increases, a hierarchical structure could be needed to accommodate the larger number of editors.

Effect of Journal Sponsorship

Journal sponsorship can influence the number of editors. For example, if a journal is one of a portfolio of journals owned by some organization, then the number of editors in the other journals could provide a reference model to be followed. In addition, journals associated with sponsorship settings ultimately could have promulgated an editorial board structure, either because of the existence of successful models from related journals or because the group effectively recommends a particular model. As a result, we would expect that journals sponsored by the same group may have similar organizational models.

Effect of Discipline

Similarly, the discipline in which the journal is based can affect the number of editors and the organizational model of the editorial board. The discipline provides a number of related journals that can be used to provide reference models of both the number of editors and the editorial model.

Further, a discipline's homogeneity or heterogeneity can influence the nature of an editorial board. For example, a highly heterogeneous discipline effectively might require editors from the particular sub-disciplines to be characterized as such in the editorial board. Hence, Management Science might denote a number of editors based on sub-disciplines.

Power and Editorial Board Size and Structure

"Power" may be a critical issue at some point in the evaluation of board size and structure. If an editorial board is smaller rather than larger, that would suggest that each member of the board has more "power" regarding the possibility of getting published in the journal. Furthermore, different editorial board structures appear to put the editorial power in different groups of editors. For example, if the editorial board uses associate editors who make editorial decisions then the power of the editor is diffused to those associate editors who have increased power. Similarly, if only the editor makes publication decisions then the editorial power is more concentrated.

Concerns with editor power probably are most likely to manifest themselves in sponsored journals. However, *a priori*, it is unclear if journal sponsorships result in diffusing publication power of the editor to associate editors who make publication decisions or if they centralize the power in editorial models where a single editor makes editorial decisions on papers.

Relationship between Size and Editorial Board Structure

Finally, it may be that editorial board size and structure are related. If there are a large number of editors it can prove to be necessary to generate organizational structure devices to help manage them.

For example, if there are a large number of editors, it could facilitate organization and processing of papers to have some editors (associate editors) charged with different decisions, such as determining whether a paper should be published. Alternatively, certain structures are likely to be able to accommodate a larger editorial board. For example, if there are associate editors charged with making editorial decisions on papers, then they are likely to need (or accommodate) additional editors to help referee the papers. Accordingly, research of these issues might also draw from the classic span of control literature.

Other Variables Potentially Affecting Editorial Boards

There likely are implicit and explicit commitments associated with being on the editorial board. On the one hand, being on a board provides status to faculty members that potentially contribute to their tenure and promotion decisions and to annual performance review decisions. On the other hand, being on the editorial board suggests that members will play a role in both providing and evaluating content. As a result, being on the editorial board of a journal can be regarded as a signal that the editorial board member "endorses" the journal. Exchange theory as developed by Blau (1964) and others could provide a theoretical basis of analysis.

Accordingly, editorial boards potentially can be used to create multiple signals. Including well-known editorial board members can be used by the journal to provide a quality signal regarding the journal content. The more well-known and well-published editorial board members listed or the "better" the universities that editorial board members are from, the better the implied quality of the journal. (Alternatively, some journals may opt for smaller editorial boards as a signal of "exclusivity"—not a likely strategy for section journals.)

Further, including editorial board members with particular research interests or from particular universities can signal that the journal is interested in specific research streams. Thus, editorial board members could provide the ability to get "buy-in" from particular academic communities, resulting in additional contributions and potential journal specialization.

In addition, if editorial board members can be expected to provide content, then the larger editorial boards potentially can provide more content and more sustained streams of content. This could facilitate journal consistency and continuity over time. Future research could examine both implicit and explicit editorial board commitments and the impact of these commitments.

Effect of Sponsorship on Time as Editor

What are the primary variables that seem to influence the editor's tenure at a particular journal? Perhaps the most important variable is likely to be whether the editorial position is for an independent journal or a society/section journal. Editorial activities for society and section journals are typically limited to "terms," for example, three or four years. Editors may get multiple terms, however, from my experience, editors rarely get more than one or two terms. On the other hand, editorial tenure for independent journals can be arbitrarily long: there are rarely specific terms of service. Although the other two independent variables (size and discipline) from Figure 9.1 may also affect editorial tenure, their overall effects are not likely as influential.

Effects of Sponsorship on "Editorial Voice"

There probably are a number of different variables that appear to influence the model of editing used by an editor. Based on my experience, one particularly important variable is whether the journal is independently sponsored.

Editing a Privately Held Journal

As an editor of an independent, privately held journal there probably is greater freedom than any other editorial environment. Generally, as long as the journal issues are produced on time with the appropriate quality, the publisher is making money, and there are no complaints, the editor faces few constraints. If the editor sees a paper that he finds interesting, he literally can ask the author to submit the paper to the journal. If the referees don't like a paper then the editor still has the prerogative to publish the paper. If an editor wants to pursue a topic she can solicit a paper on that topic.

Editing a Sponsored Journal

In some cases the journal is part of some group such as a section from a society, and the editor represents the members of that society. Editors of section journals can face political battles over the content: "There should be more x research; you are biased too much toward y research." Editorially this can mean trying to give the sponsoring member a "voice" in the editorial process. In that situation, the perspective of the members of the section becomes critical.

For me being the editor of a section journal meant ensuring that the voice of the section was heard and finding a way to embed that into the editorial process. For each paper, this meant relying heavily on referee reports that were gathered specifically from section members.

Role of Editor: Individual or Group?

Editing generally is seen as a role in a group activity, with the editor functioning within the context of an editorial board. Although it appears that most editors employ a "classical" approach to editing, there are at least two other extremes. At one extreme, the editor plays such a strong role that the entire editorial process can be reduced to a single role, the editor. At the other extreme, the editor is simply one member of the crowd and the overall approach to editing is one of crowdsourcing editing.

Classical Model of the Editor

In the classical (traditional) approach authors submit their papers to the editor who then assigns the paper to two or more referees. The editor may or may not read the paper. Each of the referees then completes a report on the paper with a recommendation that ranges from rejecting the paper to revising the paper to publishing the paper as is. After the editor has received the referee reports, she examines the consensus of the referees and makes her decision based on the referee reports.

Editing as "This Is Mine"

In contrast to the classical model, the "this is mine" editor often takes a heavy hand on most papers. For example, the editor may completely review the paper before he decides whether to send the paper to the referees. In this setting, the editor may completely rewrite sentences and even entire sections. Usually the author is not in a position to object, so most authors simply implement the editor's changes.

In the case of "this is mine" the editor makes a number of explicit suggestions that must be taken, down to the sentence level. Oftentimes the "this is mine" editor includes the referee report of a single referee. Although I have never used this model, it is my suspicion that in this case, the referee is likely to be the editor or one of his key editorial board members.

Editing as "Crowdsourcing"

At another extreme, writing and editing can take the form of crowd-sourcing. A research paper is sent to a senior editor and then to the referees and input is gathered. In general, the more referees there are, the more additional information that can be gathered. The editor may review the paper and provide an additional set of comments. In so doing the authors can get many new ideas related to the paper. Over the years I have had a number of authors thank the referees and editors for new ideas for research papers ultimately derived from the crowd-sourcing model.

Some journals push this crowdsourcing model even before the paper is submitted, encouraging the author to submit the paper to major conferences before submitting the paper to the journal. In so doing the paper gathers increasing amounts of input from the crowd. In addition, the paper gets additional credibility stamps of having been accepted and presented at the set of particular conferences.

Although the comments may not necessarily be helpful, the authors are not in a position to not accept the crowd's suggestions because the editor and referees typically will have at least another opportunity to evaluate the paper. Although some papers are heavily crowdsourced, the reader will not know of contributions other than by reviewing the acknowledgments.

Editorial Devices for Crowdsourcing

There are a number of devices that can facilitate editorial crowdsourcing. For example, papers can be posted and referees can choose to referee a paper or a part of a paper. Similarly, associate editors could choose to take control of editing a paper rather than being assigned a paper. Potentially, referees and editors could vote on whether to accept a paper. However, such approaches could arguably generate inappropriate or biased reviews from participants who are not "arm's length." Finally, there could be a concern that the editorial work might not get done: "Who will take the paper that no one wants to review?"

Which Model?

Based on my experience the classical model and its crowdsourcing nature are the most frequently used. The model in Figure 9.1 provides at least two sets of variables as to why. The use of the "this is mine" approach likely has mostly been outside sponsored society settings where the editor is based in an independent journal and there are no real "member" concerns. In addition, the number of papers submitted and processed can also influence the editorial approach. For example, if there are a substantial number of papers submitted, then the editor is not likely to have sufficient time to play a major role in evaluating each paper. Finally, the discipline also is likely to have an effect with editors in similar disciplines potentially following similar archetypes.

Evaluating Referee Reports

On the surface, being a journal editor means sending a paper to referees, waiting for their reports, and then reporting the results to the author. The ideal situation is where two or more referees each submit their referee reports in a timely manner, substantiating their evaluation with well-reasoned comments in sufficient detail so as to be able to follow the reasoning behind their recommendation, while simultaneously providing sufficient guidance to the author for revising the paper. In addition, in this ideal situation, the referees have similar judgments, of which the editor can confirm and generate a timely response to the authors. Unfortunately, the ideal situation is relatively rare. Referees oftentimes disagree with each other and the quality of their discussions of the paper may be very limited.

One Accept and One Reject

A difficult situation for editors arises when there are two referee reports, one "accept" and one "reject." In addition, to confound the situation further, in this setting, inevitably, one of the referee reports does not provide much insight into why the referee made the evaluation that he did. As a result, this setting often requires that the editor be the "tie breaker" or that additional referees be sought. In either case, this can substantially extend the review time for the paper.

Contradictory Referee Reports   While shepherding a paper through the editorial process it is not unusual to receive two widely different referee reports. In one recent case, one of the referee reports indicated that the scope of the paper should be cut down because the paper was too broad. That referee also indicated that the paper was too long. However, the other referee report suggested that the paper was too narrow and really needed to consider other issues. The second referee suggested that the paper was too short. The reviews went as far as literally to suggest excluding/including the same issues.

Although the author can try to accommodate both, it is unlikely that he will be successful in such a situation. Such divergent reports effectively require the editor to make a decision one way or the other.

The Hardest Thing Is to Say ... "Accept"

In academia, PhD students are taught to find limitations in research by analyzing a broad range of papers. Even papers that are published in well-respected journals, generating large numbers of citations, sometimes are criticized as "mistakes."

As a result, in many cases the hardest thing for a referee (or editor) is to say yes to a paper. Instead, as a referee or the editor, the "safest" strategy is just say no. After all, every paper can be improved, at least from some perspective. Accordingly, it is easy to say no to a paper because the data could be better, or the previous research is not as complete as it could be, or it could be better written, or there is more analysis that could be done. Unfortunately, we are not often taught to find the beauty in a paper.

Special (Editorial) Circumstances

As an editor there can be some "difficult" special circumstances. For example, editors may note that some contributors are particularly influential and well known (e.g., the "Big Hitter"). As another example, editors may solicit a paper from a specific user or group of users resulting in its own concerns.

Editing the "Big Hitter"

I refer to the well-published, influential, and well-known researcher, as the "Big Hitter."

An important question is, "Do editors treat papers from Big Hitters in the same way as papers that are not from Big Hitters?" Based on the behavior of junior authors who let Big Hitters on their papers, sometimes for little academic activity, it is clear that junior authors think so. Many junior authors think that all they need to do is have the Big Hitter as a coauthor and that guarantees a paper for acceptance.

Most editors employ a double-blind review process. Accordingly, the appearance of the Big Hitter as an author should be eliminated by the process. However, in some cases referees try to identify the authors. In other settings the process is not double-blind so that the reviewer knows who the authors are. As a result, there can be information corrupting the editorial process so that reviewers can determine the authors.

In my experience, the Big Hitter is not treated differently, but the question as to whether they are treated differently is an empirical one. I can say that I have found that Big Hitters typically are publication savvy. Big Hitters know better how to sell a paper and they understand basic issues such as the authors need to respond to each of the comments generated by referees.

Editing the Solicited Paper

In some settings the editor will seek out a paper from a particular author or group of authors, effectively promising to publish that paper.

Editing this paper can be problematic, because the editor wants the paper contributed but on the other hand he or she wants to make sure it is done well and meets the needs of the journal. If the paper is not initially well done it can be difficult getting the appropriate changes. In any case, soliciting papers is much easier independent of any sponsoring organizations inasmuch as there is no concern for equity of members or biasing the editorial process for a nonmember.

Editing the Special Issue

In some settings special issues of a journal are delegated to a special issue editorial team. In this case the journal editor has to decide if he trusts that editorial team "enough" or if he should provide an additional layer of editing or refereeing. Alternatively, if there are a number of special issues then that can result in adoption of the editorial board to provide the appropriate editorial participation.

What Keeps Editors Up at Night?

What are some of the major concerns of editors? Ethical issues, such as using valid data, are some of the key concerns of editors. In addition, editors are probably most concerned about getting referees actually to referee papers that they say they will referee.

Valid Data and Results

Perhaps the most important issue facing each editor relates to the information in the research papers that he ultimately publishes: Are the results real? In some cases there have been research papers where the researchers have been accused of literally making up the data.

One of the most recent cases focuses on a well-known accounting professor from Bentley University in Boston. The case of James Hunton has been documented in a sequence of newspaper articles from *The Boston Globe* (Healy, 2012, 2014) and a report from Bentley University's ethics officer (Malone, 2014, "Bentley Report").

After the retraction of a research paper because of concerns about data in the original paper (Hunton and Gold, 2013) and after the report from Bentley University's ethics officer (Malone, 2014), there was substantial concern by the sponsoring organization (American Accounting Association—AAA) that owned the journal that published the retracted paper about what the findings meant to both other authors and other research papers. For example, as noted in Malone (2014) the confidential incident reporter raised concerns about 10 other papers. Furthermore, as noted by the executive director of the AAA (Sutherland, 2014, p. 2) "... a small team has been tasked with developing and implementing the steps necessary to address the implications of the Bentley report across all journals published by the AAA." Not just the authors of the original paper would be affected, but additional authors and additional journals potentially would be affected. As noted by Sutherland (2014, p. 2) "... our next steps will involve reaching out to all coauthors of the 30 related articles published across eight of our journals, asking them to provide independent evidence of the validity of the data on which their articles are based." Accordingly, there was concern about whether some of Hunton's other previous research papers published by the AAA were based on valid or invalid data. Finally, Sutherland (2014) also noted that the AAA "... will take action as warranted by outcomes of our process." As a result, additional papers may also be affected by these investigations.

Have the Results Been Published Before?

Another critical concern is, "Has the paper been published before?" There are at least three ways that lead to determining if the paper has been previously published. First, I have had referees indicate on their reports that the paper (or one that looks a lot like the paper sent them to review) has been published before either because they knew about the previous version or ran across it as part of the referee process. Second, I have found the paper as part of an informal search to determine if the paper had been previously published. Third, there are now tools available to search the Internet to determine if the paper has been previously published or the extent to which the paper has appeared in other forms or settings.

Referees

As an editor a key activity is choosing referees for a paper. Occasionally, I have found that potential referees accept the assignment, but then do not provide a review or provide a useless review with no real substantive comments. Unfortunately, by the time that the editor finds out either situation, the paper typically has been in process for a relatively long period of time. Furthermore, typically, another referee must be found and given sufficient time to review the paper. Accordingly, I have found that the lack of appropriate referee response is the primary factor slowing the editorial process.

Summary, Contributions, and Extensions

This chapter has presented some models of editorial boards and editing and examined some of the issues that editors face. This discussion suggests that issues such as journal size (input, output, or the relationship between them), sponsorship, and discipline can have a major impact on editing and editorial boards.

Contributions

This chapter generated models of editing and the editorial board. In so doing, this chapter has generated a number of assertions about editorial boards and the editorial process that ultimately could begin to generate a "theory of the editorial process." In addition, many of those assertions and models could be analyzed empirically. For example, the chapter noted that it appears that larger journals (e.g., with more paper submissions) are more likely to employ an editorial model with associate editors who are responsible for ensuring that papers get refereed and ultimately making decisions regarding paper acceptance. As another example, the chapter analyzed the impact of journal sponsorships on a range of different variables, including editorial boards and editing.

Extensions

There are a number of extensions to the discussion in this chapter. First, the models generated in this chapter are largely based on my experience as an editor, referee, and author. However, it would seem that some existing theory could be applied to some of these issues. For example, in the models developed in this chapter, the size of the organization (e.g., number of submissions, number of pages printed, etc.) seems to affect editorial boards and editing. Similarly, organization size has played an important part in economic-based models of organizations. As a result, additional theories could be generated and embedded in these models of editing and editorial boards. Second, in this chapter, editorial board size and structure were each treated as dependent variables. However, it is likely that there is some interaction between those two variables. For example, a large editorial board likely requires more hierarchical structure than a small editorial board. Third, it can be argued that editorial boards provide a signal as to the nature and quality of the journal, a governance model of how papers will be refereed, a ready-to-use source of editing and refereeing capabilities, and a signal as to by whom and how each paper will be refereed. Further research could investigate the extent to which those signals could be captured by empirical models of editorial boards. Fourth, a number of journals employ special issues generated by either existing editorial board members or those from outside the editorial board. Typically, those special issues are based on papers from a workshop or symposium or the papers are chosen around a special topic or a limited scope. As a result, oftentimes those special issues require the ability to depend on the editors to ensure that the content is refereed appropriately. As a result, this can require an organization scheme which captures that independence. Fifth, perhaps time also plays a significant role in journal editorial boards. For example, as a journal evolves over time the structures used to govern it can evolve to include increasing levels of complexity, such as having multiple types of editorial board members.

References

Blau, P. (1964). *Exchange and Power in Social Life*. New York: John Wiley & Sons.

Healy, B. (2014). False data put all of Bentley professor's work under review. *The Boston Globe*, July 21.

Healy, B. (2012). Bentley professor resigns after his research is retracted. *The Boston Globe*, December 21.

Hunton, J. and Gold, A. (2013). Retractions: A field experiment comparing the outcomes of three fraud brainstorming procedures: Nominal group, round robin, and open discussion. *The Accounting Review*, 88(1, Jan.): 357.

Malone, J. (2014). "Report of Judith A. Malone, Bentley University Ethics Officer, Concerning Dr. James E. Hunton," https://www.bentley.edu/files/Hunton%20report%20July21.pdf, July 21, 2014.

Sutherland, T. (2014). "AAA Leader Update on Bentley University Report." E-mail, July 23, 2014.

10

THE EDITOR IS OFTEN A COACH

STEVEN R. GORDON

Contents

Introduction

Academic journal editors wear many hats: manuscript screener, review process manager, board developer, promoter, strategist, and coach, among others. Of all the hats I wore as editor-in-chief of the *Journal of Information Technology Case and Application Research* (*JITCAR*), the one that fit the best was my coaching hat. My team consisted of *JITCAR*'s editorial board, our reviewers, and our authors. Our objective: to produce the best journal possible. As a coach, I focused on being a motivator, coordinator, and developer of talent. This chapter reviews and recounts the benefits, pleasures, and rewards associated with this editorial role.

When I was first asked to be the editor-in-chief of *JITCAR*, I assumed that I would be spending all my time reading manuscripts, delegating oversight to my senior and associate editors, managing the process of assembling journal issues, and doing promotional activities. Little did I suspect that a significant portion of my time, and the time

I enjoyed the most, would be dedicated to coaching. I eventually came to believe that every editor should plan to coach his or her journal's authors as well as members of the editorial team. In this chapter, I provide support for this point of view by relating my personal experience in this role and appealing as well to theory.

This chapter is organized as follows. First, I present a brief overview of *JITCAR* focusing on its history, its authors, its editors, and its reviewers. Next, I address the many opportunities editors have for being a coach. I follow this with some observations on theories of coaching. I conclude with lessons for editors and authors and some thoughts about the joys of coaching.

About *JITCAR*

In 1999, Shailendra Palvia founded the *Journal of Information Technology Cases and Applications* (*JITCA*) to be an outlet for high-quality case-based research. At the time, the top journals in information systems and technology rarely published case-based work. Dr. Palvia envisioned *JITCA* as a vehicle that would not only provide an outlet for case researchers, but also encourage more case research by offering a dedicated space where such research would be eagerly sought. In 2005, *JITCA*'s editorial board voted to change the name of the journal to the *Journal of Information Technology Case and Application Research* so as to emphasize its focus on research cases rather than teaching cases. Nevertheless, *JITCAR* continues to publish both teaching and research cases, although teaching cases are required to have a research focus and an accompanying research note.

I edited a special issue for *JITCA* in 2002 and became one of two senior associate editors in 2004. In 2008, I accepted Dr. Palvia's offer to assume the role of editor-in-chief, a role I held for three years. During my tenure as editor-in-chief, *JITCAR* published four issues per year. Each issue contained an editorial preface, three case-based articles, a book review, and a practitioner interview called "The Expert Opinion." Of the three slots for case-based articles, two were reserved for research cases. The third article could be either a research or teaching case. Articles were accepted initially at a rate of 36%, but declined to 25% in my final year as editor-in-chief.

JITCAR was published by Ivy League Publishing, a small publisher. Although it was indexed by ABI/Inform, it enjoyed limited readership. Because of this and because only 12 articles were published each year, citation levels were low. Nevertheless, the quality of the research articles was consistently high. All articles were blind-reviewed by three members of the academic community, most of whom were selected from our editorial review board. The typical turnaround time from submission to decision was 3.5 months. Most papers that *JITCAR* eventually published went through three submission/decision rounds, with the first decision being "revise-and-resubmit," the second being "accept-with-revisions required," and the final being "accept."

As an academic journal, *JITCAR* often had to overcome the perception of authors who inherently felt that a case presents itself: just tell the story and let the readers draw their conclusions. Although that attitude might be acceptable for a journal targeted at a practitioner audience, *JITCAR* required a tighter tie to theory. *JITCAR*'s author guidelines and reviewer forms made explicit the expectation that the research *JITCAR* published would be both rigorous and relevant. Submitted manuscripts were reviewed to ensure that the research questions being investigated were clearly identified, current theory applicable to those questions was discussed, the data collection and analysis were appropriate and adequately described, and the observed outcomes were compared to those predicted by existing theory. In addition, each manuscript was assessed for its theoretical contribution in supporting, opposing, or extending existing theory and for how it might advance practice.

JITCAR's Authors

The submissions to any journal reflect a broad diversity of authors, and *JITCAR* is no exception. *JITCAR*'s relatively limited readership made it difficult to draw submissions from experienced authors, who justifiably sought a larger audience for their work. As a result, the typical submission to *JITCAR* came from an author who had little experience with academic publishing. Nevertheless, we also received submissions from experienced authors who found it difficult to publish case-based research elsewhere. Notably, among the journals now included in the "AIS Senior Scholars' Basket of Journals," only the *European Journal of Information Systems* published more than three case-based articles per year during my tenure as editor-in-chief of *JITCAR*.

Table 10.1 shows the geographic diversity of *JITCAR*'s authors. Of the 36 articles published during my tenure, the first authors of 22 were affiliated with institutions outside of the United States, most at institutions not known for their research. Of the 14 articles published by authors at US institutions, only two were written by researchers at a top-50 research university. I strongly believe, although I have no statistics to support it, that the academic institutions of authors of rejected articles would display an even weaker research pedigree. Although research excellence can sometimes be achieved at schools not known for developing research stars, researchers at these institutions often lack the support they need to turn their work into publishable manuscripts. In these situations, coaching was very much developmental, emphasizing its teaching and motivational aspects.

Table 10.1   Distribution of Articles by Country of First Author's Affiliation

COUNTRY OF FIRST AUTHOR'S AFFILIATION	NUMBER OF ARTICLES
USA	14
Denmark	3
India	3
Australia	2
Finland	2
Ireland	2
Italy	2
Germany	1
Ghana	1
Iran	1
New Zealand	1
S. Africa	1
S. Korea	1
UAE	1
UK	1
Total	36

JITCAR's Editors and Reviewers

JITCAR's editorial staff at my last issue consisted of 42 editorial board members, eight associate editors, one senior editor, two industry editors who were responsible for the "Expert Opinion" interview, a book review editor, a web editor, and five advisory editors. The size of this editorial staff might be considered too large for a journal that publishes only 12 articles per year. But, having such a large editorial staff allowed us to use editorial board members rather than ad hoc reviewers for most reviews, even though no editorial board member was required to perform more than two reviews per year. From a coaching perspective, this stable community provided the opportunity to create a sense of teamwork and to develop a unified vision and understanding of our objectives and processes, which would have been much more difficult to achieve if the team of reviewers had only limited contact with the journal.

Geographically, our editors came from 16 different countries, two in North America, seven in Europe, two in the Mid-East, two in Asia, two in Oceania/Australia, and one in Africa. This wide geographic distribution reflected the geographic dispersion of our author population. To maintain a sense of team across this geographical divide, I held online get-togethers twice a year by Skype and Dr. Palvia held a face-to-face board meeting annually at the Global Sourcing Conference. Time zone differences made it difficult for all team members to meet online in one session, so usually I held separate sessions for different parts of the world. Minutes of these sessions were distributed to all members of the editorial board to further promote a unified view of the journal's vision and to reinforce the objective of our working as a team.

JITCAR's editorial board consisted of established researchers with expertise in a variety of subdisciplines. Most members of the editorial board were also editorial board members at other journals. Our policy on ad hoc reviewers was to use well-recognized experts. To the best of my knowledge, we never used graduate students as reviewers, contrary to the policy of even some top-ranked journals.

Coaching Opportunities

Opportunities for coaching exist throughout the review process. The first opportunity arises upon the receipt of a manuscript. Unfortunately, many editors have no interest or insufficient bandwidth to assume the role of teacher or coach at this point and will desk-reject an article if it is clear that the authors need instruction. For example, at the *Quarterly Journal of Economics*, editor Larry Katz noted, "A paper will get desk rejected for sure if it is sloppy (missing references, tables and figures that can't be deciphered, equations with notation that is not defined, poor and verbose writing, etc.) regardless of the quality of the ideas" (Ozler and McKenzie, 2012). At the *American Psychologist*, outgoing editor Hyman Rodman advised incoming journal editors, "... the editor's primary function as gate-keeper—deciding which articles to admit for publication—must take precedence over his secondary function as editor. It is difficult enough to edit a journal without also running a correspondence school for would-be authors" (Rodman, 1970).

At *JITCAR* we accepted only 12 articles per year and rarely received more than 50. At this rate, I had the freedom to assume the role of coach, and generally felt obliged to do so for a significant percentage of the submissions. I desk-rejected a submission only if I felt it could not be rescued. Possible reasons for rejection might be that the paper failed to address an interesting problem, shed little new light on a topic that was already well understood, or was not based on case research. However, if the research was interesting and it was clear that the researcher had ample data to support a good paper, I would not desk-reject a paper no matter how poorly it was conceived or written. Rather than reject a manuscript with the suggestion that the authors submit it to a practitioner journal, I would try to work with its authors to advance it to a state where it could be reviewed. Technically, this was a desk-rejection, but I would urge the authors to resubmit if they felt able to follow my advice. I offered suggestions to the authors about how to restructure their work to make it acceptable to *JITCAR*. For example, if the authors had not specifically stated a research question, I would propose possible research questions that they could address. If they failed to link their research to any theory, I would suggest possible theoretical perspectives and frameworks that could help motivate and contextualize their work, and provide references for them to learn about these theories. I would sometimes describe alternative approaches they could use to add rigor to their analysis and clarity to their presentation. More often than not, the authors understood what needed to be done and resubmitted a document that met *JITCAR*'s minimum requirements to be distributed for review.

Coaching opportunities continue to arise throughout the review process. *JITCAR*'s reviewers, like reviewers at most journals, generally provided detailed explanations of what they found wanting in their notes to the authors. Often, they accompanied their objections with prescriptions about how the authors could address the problems they had identified. When reviewers failed to provide any suggestions, I tried to do so. This was a touchy process, as my expertise in the domain of the manuscript usually did not match that of the domain expert called upon to write the review. My solution was to research the issue to educate myself and then correspond with the reviewer to ask if the suggestions I had thought about might help the author respond. This process had the effect of gently, and, I hoped, in a non-threatening way, motivating the reviewer to be more prescriptive in subsequent reviews. It also reinforced the model that we were a team working toward the goal of publishing the best paper possible and established the concept that our editors were coaches as well. Finally, I copied my decision letter and all three reviews to each of the reviewers. Some other journals also make this a standard practice, but I used it as a mechanism to coach the members of our editorial team.

Theory of Coaching

A review of the literature suggests that theories on coaching are context specific. For example, a life coach, an executive coach, a weight-lifting coach, a soccer coach, and a relationship coach all use different techniques and apply different approaches, perspectives, and philosophies to their respective tasks. Nevertheless, the definitions of coaching in the literature tend to be relatively similar. Hamlin, Ellinger, and Beattie (2008) analyzed definitions of coaching in 39 different academic sources and concluded that, although there were some differences, the definitions of coaching across the categories of executive coaching, business coaching, and life coaching, coaching in general could be relatively well defined as follows: "Coaching is designed to improve existing skills, competence and performance, and to enhance [the coaching target's] personal effectiveness or personal development or personal growth" (p. 295). This definition is incomplete in my opinion, as by focusing on personal development, it ignores the organization as the coaching target. Team coaching (see Hackman and Wageman, 2005), as with sports, recognizes that an organization is more than the sum of its parts, and that personal development is just one component of team development.

Although best practices in coaching vary significantly with context, certain coaching competencies appear to arise consistently in studies (Wise and Hammack, 2011). They can be grouped into the following three areas: establishing the coaching relationship, communicating effectively, and facilitating learning and performance.

Establishment of coaching relationships in the context of journal editing isn't as simple as it might seem. Most authors see their corresponding editor as a barrier, not a coach. And, to some extent, this perception is correct, as the editor's first responsibility to the journal is to ensure the quality of accepted articles. However, both the author and the journal stand to gain if the editor and author can establish their association with each other as, in part, a coaching relationship and the author understands and accepts the editor as a coach.

A coaching relationship between the editor-in-chief and the editorial board is also difficult to achieve. Members of the editorial board are successful in their research domain and many are also successful editors at other journals. They are volunteers and see themselves as providing advice to the editor-in-chief rather than the reverse. The editor-in-chief must acknowledge their expertise, but establish the coaching relationship specifically as it applies to the mission, culture, and processes of the journal.

The communication competency is critical in both editor/author and editor/editor relationships. In the editor/author relationship, the primary components of communication are the decision letter from the editor to the author and the response, upon article revision, from the author to the editor. But, side communications are often necessary and usually beneficial for improving the manuscript and increasing the chances of success. These communications could consist of challenges to the editor's assessment and prescription, requests for clarification, and discussions about potential avenues for improvement. Unfortunately, authors are often hesitant to challenge an editor's decision letter for fear of alienating the person so responsible for the acceptance of their manuscript. They often find it less threatening and relatively easier to accept an editor's request even if they feel the request is incorrect or decreases the value of their work. They might be hesitant to ask for clarification for fear of appearing stupid. They might avoid discussing potential avenues for improvement because they are afraid of the additional work it might entail. For the most part, these fears stem from the perception, alluded to previously, that the editor is a barrier rather than a coach. Editors need to be proactive in overcoming these fears so as to open lines of communication. This requires the editor to establish, at the earliest possible point, a desire for openness and a willingness to be helpful.

It is generally much easier to achieve open lines of communication between the editor-in-chief and the other editors. Journal editors are volunteers. Most consider it an obligation and responsibility to perform editorial services for their colleagues. Editors are rarely afraid of being dismissed from their editorial responsibilities. They can always use the free time for their own research or join a different editorial board. As a result, if the editor-in-chief is willing to listen, editors are willing to speak their minds. If the editor-in-chief sets time aside for communication and deals with the logistics of creating a forum for discussion, editors are happy to communicate with one another and with the editor-in-chief. They can then look to the editor-in-chief for direction and will enjoy a sense of belonging and community.

A competency for facilitating learning is most important in the editor/author relationship. The editor's instinct, of course, is to tell the author exactly what to do to meet the editor's objections. But, corrections, in that case, become mechanical rather than the result of learning. And, the outcomes are then not necessarily ideal, as the author filters the instructions as informed by his or her prior perspective on the issue. A better approach, which may seem counterintuitive, is to be less prescriptive, highlighting the issues, explaining why they are problematical, and possibly offering multiple suggestions for addressing them. The editor should understand that individuals have different learning trajectories and that by specifying a solution, rather than encouraging exploration, learning is truncated (Doerr, 2006). We should want the author's voice, not the editor's, to be heard, otherwise the editor should be listed as a coauthor.

At *JITCAR*, the opportunity for facilitating learning often revolved around the problem of identifying a good research question. A good research question is interesting and is motivated by and related to prior research. It should be answerable by analysis of the data. For large-sample research, the research question often asks whether relationships exist between some constructs of interest. The answer is then determined through statistical testing. Case research, by contrast, rarely allows statistical testing. Instead, case research questions are generally of two types. One is exploratory, seeking to identify potential relationships among observable variables that can, at some point, be subject to large-sample study. The other builds off previously known relationships looking to enrich our understanding of them by identifying conditions or situations that enforce them, negate them, or explain the causal mechanisms behind them. The research questions associated with this type of research are often hard to formulate especially in terms of hypotheses that can be directly supported or negated by the research.

When a manuscript is rambling and the reviewers have difficulty determining where it is going, the most likely explanation is that the authors have failed to focus the case around a research question. Usually *JITCAR*'s reviewers picked up on this, but it was often my role to help the authors develop the research question in the context of the discipline.

Lessons for Editors

It should be no harder for a large journal to coach its authors than it is for a small journal to do so. The solution is for the journal to use its senior and associate editors wisely. Upon receipt of a manuscript, the managing editor, who need not be an academician, should assign it to a corresponding editor, the one responsible for communicating with the authors and selecting the reviewers. The corresponding editor could be the editor-in-chief or one of the senior or associate editors. Ideally, no corresponding editor should be assigned more than 50 articles per year, approximately one per week. At that rate, the corresponding editor should have sufficient bandwidth to coach the authors and reviewers. The corresponding editor has the responsibility for deciding whether to desk-reject a manuscript or send it for review. The editor-in-chief should randomly check rejections at this stage to determine if the corresponding editor has, for want of coaching, rejected a manuscript that meets the journal's mission and has the potential for meeting its quality standards. If the editor-in-chief determines that the corresponding editor has missed the opportunity to coach the author, he or she should open a dialogue with that editor to better understand the decision and, if appropriate, coach the editor so that better decisions can be made in the future. Even experienced editors can continue to learn through appropriate feedback at this juncture.

Lessons for Authors

All too often, upon receiving a revise-and-resubmit decision from a journal, a manuscript's authors focus entirely on what's necessary to get the paper published. But, authors should understand that they are being coached as well. Upon responding to the reviewers, they should step back and ask themselves what they've learned that can be applied more generally. If the reviewers and the corresponding editor have done a good job of coaching, the takeaway lessons should be apparent in the decision letter.

Authors should also try to open lines of communication with the editors as early and as often as possible. Even prior to submission, a letter to the editor with a brief outline of the objectives and findings of the research can be useful to ascertain if it would be of interest to the journal's readership. This can save a great deal of time in the review process. It might even solicit a "coaching" response that will improve both the manuscript quality and likelihood of acceptance, such as "Yes, we'd be interested, but we'd be even more interested if ..."

Even with rejection, authors should take advantage of any attempts the corresponding editor has made to provide some coaching. The coaching provided with rejections is often not as robust as that provided with a revise-and-submit decision, if only because the corresponding editor will not retain an ongoing relationship with the author. However, a diligent editor will understand that any coaching provided will be developmental for the author even if there is no immediate payoff to the journal. As a member of the community of researchers, a responsible editor will do his best to provide coaching even with a rejection decision.

Joy of Coaching

Aside from the productivity benefits to a journal and its authors, coaching provides emotional benefits to the coach. Coaching can and should be fun. Surprisingly, little is written on why this is the case, yet anyone who has taught or coached must surely have experienced that same intense sense of pleasure as I have when my student or "coachee" succeeded beyond all expectations. For me, the joy of coaching has many components. First, I've derived pleasure from having success-fully discharged my obligation to give back to a research community that gave so much to me in my early years as an academician. Second, I've relished the friendships I've formed at *JITCAR*, especially with the editors I worked with for the full three years of my tenure. Although I am several years removed from my role as editor-in-chief, I retain close ties with many of my former editors and even with some of my authors. Third, I've enjoyed knowing that I have affected the research trajectory of so many of *JITCAR*'s authors in a positive way. I continue to follow their success in publishing subsequent research. Finally, as I look at my shelf containing the journal issues I was so intricately involved with, I am happy with pride in my contribution to the output.

I would be remiss if I were to give the impression that being an editor is nothing but fun. It is a tremendous amount of work and there are disappointments and perils along the way. Not all coaching is successful. Although I know intellectually that must be the case, coaching failures always felt personal to me, especially when promis-ing papers were withdrawn or, never having met their potential, were rejected. I always felt that had I done a better job, the results could have been positive. Time stress was also an issue. As the publication date neared, time became compressed, and tens of problems needed to be solved immediately. Despite all this, my memories of the job are overwhelmingly happy ones. I believe that anyone who likes to coach will enjoy being an editor.

References

Doerr, H. (2006). Examining the tasks of teaching when using students' math-ematical thinking. *Educational Studies in Mathematics*, 26(1): 3–24.

Hackman, J.R. and Wageman, R. (2005). A theory of team coaching. *Academy of Management*, 30(2): 269–287.

Hamlin, R.G., Ellinger, A.D., and Beattie, R.S. (2008). The emergent 'coaching industry': A wake-up call for HRD professionals. *Human Resource Development International*, 11(3): 287–305.

Ozler, B. and McKenzie, D. (2012). *Q&A with Larry Katz, editor of QJE*, Jan 4, 2012, http://blogs.worldbank.org/impactevaluations/qa-with-larry-katz-editor-of-qje, accessed on June 23, 2014.

Rodman, H. (1970). Notes to an incoming journal editor. *American Psychologist*, 25(3): 269–273.

Wise, D. and Hammack, M. (2011). Leadership coaching: Coaching competencies and best practices. *Journal of School Leadership*, 21(3): 449–477.

11

HAPPY MARRIAGE OR ODD COUPLE

Reflections on Editing the *Journal of Homeland Security and Emergency Management*

IRMAK RENDA-TANALI AND SIBEL MCGEE

Contents

Introduction

The *Journal of Homeland Security and Emergency Management* is an online journal that was initiated in 2004. Emphasizing the complex linkages between the two fields in its title, this journal has aimed to address a critical academic and operational need: contributing to strengthening of the much needed dialogue, collaboration, and knowledge-sharing between the communities associated with both fields. Over the years, a few more similarly focused journals have emerged, while the overall number of online published journals skyrocketed. However, when the *Journal of Homeland Security and*

Emergency Management was initiated, it was the first of its kind, in terms of content, delivery style, and platform. We experienced quite a few stumbling blocks during the journey of establishing the journal as the premier journal in its domain. It has been particularly challenging to keep up with the continuously evolving perspectives of the practitioners and academic community—the target audience of our journal. With the increasing number of large-scale disasters and emerging threats in recent times, not only have the safety of our nation and the well-being of the rest of the world been challenged, but also our thinking paradigms have been altered. We have found managing a journal that addressed such dynamic and complex issues that are under constant reinvention and reformulation to be a formidable challenge. With this chapter, we share some of the smart practices we discovered during the process that could aid new editors as well as inform anyone who is interested in learning some of the intricacies of the world of academic publishing. After providing a brief history of our journal, we present the lessons from our experiential journey through specific headings so that the readers can review the content quickly and focus on issues that may interest them the most.

History and Background

The *Journal of Homeland Security and Emergency Management* (*JHSEM*) was first introduced by two prominent disaster research scholars about a decade ago. Both were serving as faculty in graduate school, conducting cutting-edge research and providing proactive consulting, dealing particularly with interdisciplinary aspects of how governments, nongovernmental organizations, and communities leverage the systems, processes, and technologies for disaster mitigation, response, and management. Due to the dynamic nature of the field, textbooks either did not adequately address the emerging issues or were outdated in the face of new crises, emergencies, and security threats. Particularly, the public outrage that followed the 9/11 terrorist attacks led to a plethora of changes in how crises and disasters were to be conceptualized and addressed.

Before the 9/11 attacks, the possibility of any major terrorist attack on American soil was perceived to be unlikely, if not impossible. Although there were previous terrorist attacks, they were different from the 9/11 attacks. The Oklahoma City bombing turned out to be an isolated act by one individual, and the 1993 bombing of the Twin Towers in New York City had faded in memories as the perpetrators were already identified and brought to justice. With the 9/11 attacks, the national emergency management focus suddenly shifted from preparedness for natural disasters (e.g., hurricanes, earthquakes, tornados, floods, etc.) to man-made disasters, particularly terrorist attacks, which would potentially have greater economic and social ramifications for the Nation and across the globe. The educators and researchers of emergency management had to rethink their paradigms to include the anticipation and mitigation of, preparedness for, and response to all hazards, and not just the natural disasters, that would jeopardize the security of the homeland.

Subsequently, major changes in legislation led to the adoption of a new doctrine of homeland security, which informed and organized the federal, state, and local level emergency management activities in significantly different ways. This also meant that the existing curricula, textbooks, and research programs had to be revised. The number of universities offering emergency management or homeland security curricula or programs dedicated to the respective fields has increased exponentially. Even though a significant amount of funds were beginning to be invested in researching the root causes of emerging threats (particularly radical Islamist terrorism), it would take months and years to see those research findings to be translated into journal articles that would allow access by a wide range of audience. Against this background, *JHSEM*'s founding scholars, thanks to their foresight and commitment to professional and public service, recognized the need for a dynamic and scholarly publication venue that would address leading homeland security and emergency management issues. The new journal that they initiated presented a novel approach, combining the two fluid and complex fields (i.e., homeland security and emergency management) and quickly disseminating the most recent research findings and information to students and practitioners of these fields.

Homeland security is now a known field and refers to the protection of the people of the United States in their homeland and the Nation's critical infrastructures and key assets from major disruptions, be it through a major natural disaster, a terrorist attack, or an accidental/industrial threat. *JHSEM* serves as the premier publishing venue that emphasizes a holistic approach, interdisciplinary perspective, and cross-communication between academic and practitioner communities. The mission of *JHSEM* as it stands today is highlighted in Box 11.1.

BOX 11.1   *JHSEM*'S MISSION

JHSEM promotes a comprehensive and dynamic perspective, providing readers with up-to-date information regarding the evolving nature of the homeland security and emergency management fields. Recognizing the inherent links between these two fields, the journal aims to serve as a bridge between them, encouraging exploration of their underlying relationships, interactions, and synergies.

JHSEM's mission is propelled by the conviction that in fields that share significant operational elements, such as homeland security and emergency management, both researchers' and practitioners' insights are needed for success; in isolation, their perspectives can offer only partial views of the reality. Accordingly, *JHSEM* also aims to serve as a bridge between the researcher and practitioner communities, facilitating the kind of collaboration and coordination that enables them to appreciate and benefit from the associated implications and inherent connections in their fields. It is only from the intersection of these two perspectives and the interchange of ideas that the most compelling knowledge, novel insights, and best practices can emerge.

A critical part of *JHSEM*'s mission is to enrich the perspectives of researchers and practitioners in the homeland security and emergency management fields so that they are better able to address the increasingly complex issues before them. To this end, *JHSEM* encourages an interdisciplinary approach that reflects the expanding boundaries of these two disciplines by including such content as public health, cyber security, and environmental policy.

Source: http://www.degruyter.com/jhsem

BOX 11.2   *JHSEM*'S OBJECTIVES

- Meet the need for peer-reviewed, high-quality, wide-ranging professional articles in the fields of homeland security and emergency management.
- Augment the work of existing professional societies and single-discipline publications in these fields.
- Follow developments in the two fields and offer new information and analyses as they become available.
- Serve the needs of both the academic and practitioner communities by providing a high-quality platform for information-sharing, collaboration, and exchange of ideas in an efficient and economical way.

Source: http://www.degruyter.com/jhsem

JHSEM quickly became known across the respective communities. Both scholars engaged in the research of homeland security and emergency management fields and the professionals and practitioners who were tasked to formulate and implement relevant policies took interest in the readership as well as the authorship of the journal. The key objectives of *JHSEM* which were formulated back then and which stand true as of today are summarized in Box 11.2. Even though the editorial board and the editors of *JHSEM* may have changed over the years, *JHSEM* continued to pursue these objectives.

Below, we compiled some of the lessons we learned as we strive to keep *JHSEM* as a successful and relevant resource for both academics and practitioners of the two dynamic and interrelated fields. Although the details in each section are specific to *JHSEM*, we hope that other editors will find the underlying insights applicable to their own fields.

Keeping Focus

Not losing sight of a journal's mission and vision in the face of temporal changes and editorial turnovers may prove more difficult than expected. As a result, a journal may find itself fulfilling a function different from its original emphasis due to a variety of reasons, ranging from editorial preferences to incomplete or inadequate articulation of its mission. For many years, *JHSEM* has served as a barometer in the fields of homeland security and emergency management. In fact, a review of the articles published in *JHSEM* would allow one to gain a quick understanding of the evolution of these fields over the course of the past decade and the shift in focus as a reaction to actual events (e.g., Hurricane Katrina, H1N1 influenza virus, the shoe bomber terrorist plot, Hurricane Sandy, etc.).

However, until about two years ago (and before the current editorial team), the journal had evolved to publish what came along its way without much strategic thinking, and its focus shifted from publishing original research articles to an emphasis on book reviews. The latter served the community in useful but nonetheless different ways from what *JHSEM*'s original mission statement envisioned. In addition to deviation from its key objective, *JHSEM*'s featuring of an overwhelming number of book reviews inadvertently led to questioning of *JHSEM*'s standing as a high-quality publication platform. Emergence of competing (many online) journals provided alternative venues to those researchers who have been confused by the large numbers of book review articles published by *JHSEM*. Moreover, the impact factor, a key performance measurement indicator for a journal, did not take into account book review articles. As such, spending time, energy, and resources on review articles did not contribute to the journal's mission.

JHSEM needed to be put back on the forefront if it were to survive. Upon assumption of responsibility, the current editorial team reevaluated *JHSEM*'s vision and decided that the original aim and scope needed to be restored. The number of book review articles was reduced and the emphasis was once again placed on original research articles. Robust understanding and commitment to a journal's mission is a key requirement for editors to keep their journal on track and meet their audience's expectations, thereby ensuring their journal's long-term viability.

Achieving a Harmonious Balance between Competing Tendencies

A key challenge for journal editors in general may be to find a way to strike an acceptable balance between those aspects that lead to tension in their domains. In the case of *JHSEM*, editors have strived to accommodate two competing perspectives: philosophical and practical. The homeland security and emergency management fields have evolved over the years into a more integrated paradigm; a complex web of relationships and interactions between stakeholders, processes, capabilities, and missions of the two fields make studying/researching these two fields in isolation not only difficult, but also less desirable. Yet, opinions vary regarding whether the homeland security and emergency management fields should be combined either in concept or in practice. The tension between the two camps that are for and against merging of the two fields manifests itself in different contexts ranging from discussions of whether the Federal Emergency Management Agency (FEMA) should remain within the Department of Homeland Security (DHS) to whether the degree programs should offer a single diploma, demonstrating qualifications in both fields.

Conflicting views of what homeland security and emergency management are and how one relates to the other have not helped to establish a clear consensus on the necessity of multiple viewpoints. In fact, these two fields lie in the intersection of many disciplines and as such would benefit from leveraging insights from all of them to include, among others, public administration, medicine, operations research, management, political science, psychology, sociology, engineering, and information technologies. The requirement for a multidisciplinary approach has many implications, one of which is the necessity to bridge the academic and practitioner perspectives. *JHSEM* editors subscribe to the idea that the most accurate, complete, and rich insights will emerge from the synthesis of knowledge and experience of these two groups (i.e., academics and practitioners). Instead of choosing one of these two camps as our target audience, we decided to facilitate collaboration and exchange of ideas between these two groups. As such, *JHSEM* encourages and publishes submissions by both groups, hoping to turn this tension into a constructive harmony that will enhance both fields' academic and operational success. Enabling synthesis of insights and capabilities of both the academic world and practitioner communities may prove useful for other journals that address fields with a strong operational element. The underlying lesson goes beyond bridging the gap between academia and practitioners. Editors may indeed find identifying other tensions in their fields and developing ways of reconciling conflicting perspectives equally rewarding.

Tactical Challenges

Maintaining Quality

Maintaining quality standards is the surest way to ensure viability and relevance of a journal. Journal editors are better served when they do not sacrifice quality under any circumstances. Ensuring high-quality standards for a journal becomes easier if editors formulate and comply with a well-developed and transparent decision-making process. Such a process not only provides a solid rationale for the decisions rendered, but also depersonalizes negative assessments by bringing the focus to the quality of the actual submission rather than the quality of the submitting author. This minimizes the potential tension between the editor(s) and the author(s) in the case of an unfavorable decision and the subsequent appeals.

Establishing a rigorous and blind peer-review process with the help of clear and concrete guidelines for manuscript evaluation is a critical part of a conflict-free, transparent, and informed decision-making process. The role of associate editors and reviewers in implementing the peer-review process cannot be overstated as they ensure low-quality manuscripts are rejected and promising manuscripts are improved to sustain the journal's quality standards. With a well-substantiated (*JHSEM* relies on reports from three reviewers) and blind peer-review process, both the editors and the reviewers will be better equipped to eliminate low-quality manuscripts even if they are authored by well-known and otherwise successful scholars/practitioners.

Perhaps less clear is the role of prefiltering (undertaken by the managing editor). Prefiltering allows submissions that have been deemed not ready for peer review to be rejected without a peer review. This preliminary evaluation is helpful to curb the undue workload both for assistant editors and reviewers. Assuming all submissions should have their day with the reviewers not only disrespects the time and voluntary services of the reviewers, but also risks lowering the quality standards of the journal. Many submissions are far from being ready for peer review and prefiltering can be a crucial strategy to eliminate them without investment of scarce resources.

Another smart practice in journal management is that of maintaining two separate lead editorial positions with clear roles and responsibilities. In the case of *JHSEM*, the editor-in-chief is tasked with strategic direction and vision of the journal to include networking with professional organizations, promotion of the journal, and recruitment of authors and reviewers. The managing editor, on the other hand, is responsible for day-to-day operation of the journal to include prescreening, providing continuous guidance and mentoring to associate editors, assignment of manuscripts to associate editors, adjudication of conflicting reviewer positions, when necessary, and communicating with authors about undesirable but sometimes inevitable cases of mutual frustration. The presence of two lead editors with different roles and responsibilities serves to ensure a manageable workload for both editors. Moreover, in cases of conflict of interest (cases where the managing editor knows the author or has an affiliation with an organization), the editor-in-chief can serve as acting managing editor to ensure a fair peer-review process. An additional benefit from this team approach is the insulation of the editor-in-chief, who cultivates the journal's network, from possible adverse effects of negative decisions rendered by the managing editor.

Keeping Your Journal Rich

Maintaining your journal's attractiveness to its audience requires its content to align closely with developments in the subject field. Keeping a wide range of submission categories is one way to attract different types of research. Manuscript type alone, however, cannot guarantee comprehensive and relevant scope for a journal. For the latter, editors need to commit to capturing and bridging different perspectives as well as close monitoring of emerging needs and trends in their fields. In the case of *JHSEM*, we found it helpful to have manuscripts in the categories of research article, news/communique, and opinion. As discussed earlier, we also seek to diversify *JHSEM*'s content by encouraging and accepting submissions from both students and practitioners of the homeland security and emergency management fields. These two groups of authors clearly provide different but complementary insights that enhance our understanding of the related systems, key issues and processes, and the related problems.

JHSEM also welcomes research and analysis of homeland security and emergency management-related issues from authors both within the United States and other countries. Increasing interdependence, globalization, and growing complex connections across the globe make all nations vulnerable to similar risks and adverse events. Intrinsically, applicability of research conducted in one country to another that is concerned with similar threats and challenges comes as no surprise. Similarly, featuring research that leverages different disciplines offers new angles that otherwise may be difficult to attain through traditional approaches. As such, being open-minded about submissions that may appear tangential to or distant from a journal's core domain may at times prove fruitful for attaining novel insights and nonintuitive findings. Instead of rushing to reject, faced with a promising manuscript, editors can condition acceptance to a strong and explicit discussion by the author of implications of the findings for the journal's specific field or other contexts (e.g., other countries).

Special Issues

In addition to maintaining variety in submission categories, submitting authors, and disciplinary approach, topical diversity is critical to keep different audiences engaged in a journal's content. Such diversity may be accomplished by periodically evaluating published content and assessing gaps in the light of relevant and emerging issues. For utmost situational awareness, editors should continuously monitor trends in both their journal's topical domain and content published by competing journals.

A great way to fill gaps in subjects featured in a journal is to publish at least one or two special issues addressing emerging topics from a variety of perspectives. The first and the most crucial step in initiating a special issue is to identify a well-known, capable, and committed subject matter expert who can oversee a special issue. Among the adjectives we used, "capable" and "committed" are the most important ones. Special issue editors should be resourceful in gathering potential manuscripts. Sending mass e-mails to the professionals and academics who engage in that particular subject area is unfortunately not sufficient. Special issue editors need to be particularly proactive and committed with willingness to leverage their social skills and networking abilities to identify and follow up with key contributions. If editors think it is enough for a special issue editor to be a subject matter expert and write a good call for papers announcement, they will be frustrated to see that the number of viable submissions may not exceed the number of fingers of one hand. Canceling special issues can be seriously embarrassing and may hurt the image of the journal.

Communication, Coordination, and Management Skills

Journal editing is a difficult job that is not meant for everyone and requires the same key traits that successful managers have in common, among others, good communication and social skills.

As editors, it is important to maintain courtesy and respect toward authors (be they novice graduate students or worldwide famous scholars), your associate editors, and other colleagues (reviewers, producers, copy editors, etc.), all of whom ensure your journal's operation. Commitment to timeless social and professional etiquette can go a long way. Because most of the journal management-related communication takes place in writing (particularly via e-mail), editors would benefit from following simple yet powerful principles of correspondence. For example, avoid generic correspondence (e.g., "dear author" or "dear reviewer"). Personalizing decision or request letters with a greeting and the actual author/reviewer last names (first names can be used if you know them personally) can increase perception of approachability and authors'/reviewers' desire to work with you. We also recommend using their title with last names and erring on the side of referring to authors/reviewers as "Dr." In our experience, individuals mind declining a generously granted title less than not getting appropriate acknowledgment of their status.

Finally, even when rejecting a poor submission, it is wise to thank the authors and encourage them for potential future submissions. In cases in which you find the submission topic out of your journal's scope, suggesting alternative publication venues will help you maintain good relations with authors and potentially encourage them to revisit your journal with a more appropriate submission in the future.

Ethics and Accountability

It is not uncommon for editors to be caught in a situation where reviewers present conflicting positions. Conflicting reviews do not necessarily mean one or more reviewer(s) did not apply due diligence to their evaluation. Reviewers bring different perspectives and levels/ areas of expertise to the peer-review process. For example, whereas one reviewer has deep expertise on the topical area, another reviewer may focus more on the methodological issues than the subject matter itself. In such cases, the associate editor (or the managing editor if the associate editor requested adjudication assistance) has a few options. After reviewing the manuscript and reviewer reports, you can contact reviewers for further elaboration if you disagree with their critique. Alternatively, you can use your own judgment to render a decision, being well prepared to present your reasoning. You can also consult with other editors (including the editor-in-chief) or send the manuscript to a third (fourth or even fifth) reviewer. Generally speaking, we would recommend going with the latter option (consultation with other editors and new reviewers) in the case of controversial topics or low-quality reviewer assessments. This may cause some delay in your processing; however, turnaround time should not override quality and accountability standards.

Journal reputation is extremely important and is vulnerable to both objective performance measures (e.g., impact factor, turnaround time, etc.) and subjective evaluations (e.g., inaccurate rumors or unfair stories). As such, accountability is critical for maintaining the good standing of your journal's reputation. Editors should always document the rationale for their decisions and present feedback that is free of subjective expressions (e.g., "I feel like" or "it seems to me that"). A decision letter that clearly articulates objective flaws with a manuscript (e.g., issues with research design, methodological approach, factual errors, inadequate substantiation, logical fallacies, inadequate or poor analysis, etc.) and backed with reviewer reports is less likely to be appealed.

Ethical concerns should also be in the forefront of editors' minds to keep their journal's reputation intact. Requests for preferential treatment (e.g., faster processing time, acceptance of a manuscript, etc.) should be denied and in cases of conflict of interest, alternative editors should be designated for manuscript processing. Editors must avoid any situation that may damage the journal's ethical standards by making it clear to the authors that the peer-review process cannot be bypassed and manuscripts (even those manuscripts that are already revised and resubmitted) cannot be guaranteed acceptance.

Maintaining fairness and quality is as important as quick turnaround time.

Finally, the lead editor team (editor-in-chief and managing editor) must maintain a collegial working environment and set the example for trustworthy and honest relationships for the associate editors as well as the reviewers. The character and work ethic of the editors inform and condition how the journal operates and is, in turn, perceived by others. Even under the most frustrating circumstances, it is important to maintain cordiality and professionalism in journal management.

Keeping Track of Performance Measures

Management by objectives (MBO) is one of the key elements of journal editing, if you want to ensure your journal's viability and accomplish its expected outcomes. There are several objectives that are common to almost all journals (e.g., publishing high-quality articles that align with the mission and objectives of the journal, meeting the requirements of the publisher, closely monitoring publication quality and performance measures, increasing attractiveness of the journal, gaining new readers, etc.). Yet, even the seemingly most obvious objectives may prove challenging to attain in the face of persisting disagreements.

For example, journals still operate as a form of business and their survival depends on their publishers' support and marketing. As such, journal editors need to work with their publishers to ensure they fulfill the publishers' expectations. Publishers often have requirements in terms of number of articles and issues per year and possibly subject areas to be covered. Publishers' requirements may not always align with the editors' vision and finding an acceptable balance between what the publishers want from the editors and what the editors believe is feasible requires clear communication and successful expectation management.

Another controversial yet common journal objective is concerned with attaining superior performance measures. Impact factor is a numerical indicator of the relevancy of the journal to the community. Its calculation relies on the proportion of the citations in a given year out of the number of previous year's published articles. Yet, the impact factor is dependent on a variety of factors, some of which may be beyond the editors' control. Among these are the age of a journal (journals that have been in existence for a long time have an advantage over those that are newly established), state of the subject area as an established field, library indexes in which the journal is included, the number of competing journals, marketing strategy, ease of access (e.g., online delivery, print copy, both), and subscription rates. For example, if your journal is competing against relatively more widely known and established journals, its impact factor is likely to be low.

The impact factor is a tricky figure and whether it can represent or measure a journal's true success is a topic for ongoing discussions. Those who question the validity of the impact factor as a true metric of journal performance point to those indicators that factor calculations do not take into account. "What is your journal's turnaround timeline?" is a question authors frequently ask editors. Quick turnaround timelines may imply a less rigorous quality control process whereas slow turnaround timelines may mean that the articles are not attended in a timely fashion. Because the turnaround timeline often has implications in terms of timely publication of research findings and tenure decisions of faculty, authors are likely to avoid journals that they know will not process their submission in a timely fashion. As such, turnaround time is a significant performance measure for a journal, one that is not included in calculations of the impact factor. Similarly, journals that target practitioners may be unfairly judged as practitioners' use of a journal is not easily documented through citations in a publication.

Notwithstanding the controversy about whether the impact factor is useful to determine a journal's success, editors may choose to focus on increasing their journal's impact factor as a strategic goal. This is a thorny objective. If you publish too many articles, then the number of citations out of those articles has to be equally high. Yet, if you publish too few articles, the journal may fail to be comprehensive, attracting only few researchers in the field. Some editors may choose to recruit well-known authors hoping their articles will result in a large number of citations. However, such a strategy may hurt the diversity in topics and perspectives represented in the journal. A much healthier approach may be to recognize potential limitations of impact factor and avoid making it the focus of your journal editing. Editors would be well served by identifying and monitoring multiple (alternative) measures of performance to gain a complete and accurate assessment of their journal's state.

Conclusions

Journal editing is a long and tedious process. Although "journal editing" is not a money-making business, it brings significant prestige and credentials to editors in their fields. However, the issues we described above and many more that we skipped due to space constraints show that it is not for everyone. In fact, we would recommend editor positions to only those who have great passion for their field and whose vision goes beyond their self-promotion and includes serving a greater community. The job is suited for those who believe they can make the world a better place by helping create, share, and disseminate information and knowledge.

As a closing note, we would like to encourage ambitious editors to embrace a crucial mission. Information and knowledge are most relevant and useful if they have practical applicability to make our lives easier, better, and safer. As such, we believe that publication platforms should not be biased toward featuring only articles by academics or practitioners. Particularly in fields that have strong public policy and operational elements, academic and practitioner perspectives complement each other. Information-sharing, exchange of ideas, and shared understanding between these two groups of stakeholders can increase the odds of resolution of the most complex problems we face. We need more platforms facilitating this dialogue. Unfortunately, in the publishing world, and even in fields such as emergency management and homeland security, there appears to be a prejudice toward journals that aim to synthesize operational and academic perspectives. The academic community should be more receptive to the idea that journals capturing operational issues have valuable insights to contribute to the development of the field. Similarly, practitioners should be more receptive to academic research rather than dismissing key findings as theoretical discussions with no real-life implications. Editors who are committed to bringing these two perspectives together are likely to oversee journals that generate most novel and useful insights.

12
Rules for Referees

DENNIS E. LOGUE

Contents

If you have been asked by the editor of a peer-reviewed journal to referee a paper, you should be honored. That you were asked is testimony that you have risen to a position of some prominence in your field. If the editor did not believe that you could do the job, you would not have been asked. This is one price you pay for prominence.

If you are an experienced referee, then what I have written might be properly viewed as reminders. If you are a novice, I hope this helps in forming your approach to refereeing.

Let me be more specific. One reason you may have been asked is that you are an associate editor of the journal. You have achieved a position of honor in your field. The presumption is that you are capable of doing the job, and that you are able to handle the task carefully and in a timely manner. Suppose you lack sufficient expertise to referee the paper. Your own past work touches only peripherally on the topic about which you're being asked to comment. In this case, send the paper back to the editor as quickly as possible (within days, if snail mail; in minutes, if e-mail) so that the editor can reassign the paper. All authors, especially newly minted PhDs are anxious for feedback.

Another reason you might have been chosen as a referee is your own past work. When I edited a journal, I used the bibliography (reference list) as a guide to whom I might choose to review a paper. Most, if not all, editors I know or have known use the bibliography of the paper to help identify prospective referees. There is the presumption that someone whose work is extensively cited actually knows something about the topic. Accordingly, a potential referee may have received a number of citations. (Indeed, this is a trick authors might use to channel a paper to someone known to be sympathetic.)

A researcher may, however, be cited for making important theoretical contributions, but the paper is heavily empirical, perhaps sufficiently so that a dedicated statistician or econometrician is necessary to assess the paper properly. Alternatively, you may discover that your work has been cited because the new work contradicts your own. Best to leave your bias alone and return the paper to the editor for reassignment. Conflicts of interest must be avoided. If after reading the abstract and skimming the paper's introduction and conclusions, you conclude that the paper falls outside your expertise, let the editor know as quickly as you can.

A third reason you might have been chosen is you are a young scholar who has done a little work in this topic. In this case, perhaps you are being auditioned for more refereeing and perhaps an associate editor position. You are flattered. You really want to do a good job, but again the paper may not be quite "in your wheelhouse."

You can take three paths after reading the abstract and skimming the paper and discovering it lies outside your expertise. You can undertake to study up on the topic, taking many weeks to get caught up on the literature so that you can place the paper's quality in the context of previously published work. For young scholars, this may not be a good idea. It will take time away from your own work. (So do not follow this path unless you really become intrigued by the topic or at least after an assistant editor appointment.)

The second path is returning the paper to the editor with a note explaining that you do not know enough about this topic or methodology to review the paper properly.

The third path is faking it. You obviously must have some familiarity with some aspect of the paper or else you would never have been sent it. So you focus on the parts you know something about—for example, the theory, the empirical tests, the broad conclusions—thus you review a fraction of the paper thoroughly and make some general comments about the rest. Doing this, you may "miss the forest for the trees." The author may have made a meaningful theoretical contribution which you miss, but you attack the methodology which is satisfactory but perhaps not as robust as possible. Because you want to impress the editor as a capable and speedy reviewer so you can be considered for an associate editor's position, you may cause a meaningful paper to be turned down or a horrible paper to be accepted. Either way, you have not helped the author or the editor.

I, personally, have been victimized by such a referee. A very distinguished coauthor (economist) and I authored a clinical economics paper showing that the Government Sponsored Enterprise (GSE) Tennessee Valley Authority (TVA) could never bring its debt down to the levels it forecast in its own published business plan without major changes in the way it operates or is financed. The piece was turned down by a referee who contended that the TVA could be brought to heel if it had to rely on short-term debt and face the prospect of not being able to refinance. The referee missed the point: the TVA was GSE, so could always refinance because of the implicit government guarantee. There were many other features of the paper but this was key. The TVA would always be financially unsound.

As noted, it was applied research and sent to a journal that published such clinical papers. It was rejected there, but subsequently published. The "kicker," however, is it became the subject of a feature article in *Barron's* magazine, a far more widely read publication than the combined readership of both academic journals, the target where it was rejected and the place it was eventually published.

I also call your attention to an interesting paper by Gans and Shepard (1994). They provide a long list of very important papers in economics and finance that were rejected. Sometimes after the author fought with the editor, the editor asked for a fresh review, and sometimes the editor stood firm. Fortunately, these papers found homes, and the field of economics is the richer for it. What we will never know is whether the initial rejections were the result of lack of expertise by the initial referee, bias because of conflicts with the reviewer's own work, or "faking it." As you should recall from basic statistics, these authors were the victims of Type I errors. Although they cause authors anguish, they may not be deadly. Good papers can always find a home.

Hence, perhaps the most crucial rules: know the subject matter well, perform the referee's duties in a timely fashion, or return the paper to the editor and give the editor a clear impression of its quality and fit. You can do these on the basis of a quick read of the abstract and the beginning and ending of the paper.

As John Lynch (1998) reminds us, naturally all reviewers are authors and many authors become reviewers. A referee is both critic and coach. As authors, we crave helpful and prompt reviews. We want the reviewer to be careful and perhaps insightful. We are especially hopeful that the reviewer, unless accepting the paper unconditionally, will help us improve the paper, possibly turning a minor contribution into a major one.

Alas, as reviewers, our interests are not as closely aligned with those of the author. If a reviewer rejects a paper that subsequently gets published and acclaimed, only the reviewer knows for sure of the issue. Even the journal's editor may not recall a clunker of a review. Furthermore, if a reviewer helps a mediocre paper become a significant contribution, again only the reviewer knows of this important deed. I know of several journals that revealed the names of reviewers in a footnote to the published papers, so the authors and other readers could make judgments of the reviewer; this is at least a step toward reviewer accountability. Unfortunately, these journals no longer follow that practice.

Reviewers operate in obscurity with very little public accountability. Accordingly, reviewing is truly one task where "virtue is its own reward" is supposed to provide all the motivation necessary. As far as I know, there is one small exception. Ivo Welch (2013) edits the *Critical Finance Review* and publishes a blog. At the blog's end, he makes public the names of reviewers who accept the assignment but then fail to do so. This practice introduces at least a modicum of accountability to the referees.

Alas, there is no formal rebuke of referees who make Type II errors. Accepting a paper that is wrong, or even below the standards of the journal, carries with it significant costs. Authors of better papers get stuck in the queue, forced to wait extra months or even years before seeing their papers in print. Of course, during that waiting period, their ideas may be scooped. In addition, there is a cost to readers. They may read a paper under the belief that it has been carefully appraised only to find out later that it was misleading or seriously flawed and wrong. The careless referee has cost them hours as readers and maybe days or weeks if they try to build on the faulty research. This is something experienced referees and novices, as well, should strive to avoid. You are not doing anyone a favor by accepting a substandard paper; not even the author gains because peers will discover embarrassing mistakes. If these peers become gatekeepers for the authors' promotions, job opportunities, or awards, a bad paper can hurt very badly.

Finally, note that the referee has an obligation not only to the author, but also to the journal's editor. The editor deserves timely, carefully crafted referee reports and a clear recommendation.

Starting the Review[1]

After deciding that a paper can be reviewed with expertise and in a timely manner, say six to eight weeks, you should read through the paper quickly to get a feel for the landscape, then read it with great care.

Pat Thomson (2012) has created a list of questions that the careful reader should ask. I reproduce here only the lead questions in each sequence. I encourage you to go to the source. Indeed, every serious referee really ought to refer to the complete article. The lead questions are:

(1) Does the paper fit the journal?
(2) What is it about the paper that will be of interest to readers of this journal?
(3) Does it establish a clear warrant for its topic within current policy/practice of the journal?
(4) Does it have a point to make?
(5) Does it refer economically to the key literature and/or theoretical resources it needs to make its case?
(6) If it is an empirical piece of work, do you know enough about how the research was conducted to trust it?
(7) If it is a theoretical piece, is there sufficient detail about the theory to allow you to follow the way it is used?
(8) If it is an empirical piece of work, is this reported in a way that is comprehensible and defensible?
(9) Does the conclusion address the so-what question? (Does the conclusion make you glad you read the piece?)

(10) Is the abstract a fair representation of the article?

(11) Is the article well written?

(12) Does the article meet the journal conventions in titling, headings, referencing and word count?

Each of these leading questions has many subparts, so I direct you to the Pat Thompson (2012) blog.

In my mind, these are not all equally important, but all are quite important. Questions (1) and (2) are very important. The paper should match the journal. The referee should be intricately familiar with the journal's mission statement. If the paper is directed toward theoretical researchers, if published in an empirical or practitioner-oriented journal, it will be neglected perhaps despite great insights. So if a referee discovers this, the reviewer should write a note to the editor that the author should try journal *X* where the audience may be a more appropriate one.

Questions (4) and (9) should be obviously important. Without a point, what is it? An essay on how I spent my summer vacation?

Clearly (6) through (8) are at the heart of the referee's task. These establish whether the author is right. If they cannot be answered positively, the paper goes back either for massive revision or rejection.

Questions (5), (10), (11), and (12) are issues that can be remedied with careful editing, assuming everything else works, although many authors may have trouble with (11). Indeed in economics I can think of several Nobel laureates and near laureates who are absolutely horrible writers. Fortunately, the strength of their ideas got their referees and subsequent readers to struggle through their papers.

Evaluation and Recommendations

The author has spent many hours preparing the paper. (Indeed the time spent on a doctoral dissertation which, with luck, will serve up at least one publishable paper might be several hundred hours.) Whether the paper is good or bad from the perspective of the reviewer, it deserves respect. If it is great, immediate acceptance may be warranted. If it is all right, perhaps it can be fixed or redirected to a different journal. If it is really poor relative to the target journal, it should be rejected, or perhaps redirected to a more suitable journal. Even the poorest paper deserves some commentary about why it is not suitable for publication in the target journal.

I had a colleague who nearly left the profession after his first submitted paper received a review that began, "All copies of this paper should be burnt." It was followed by an explanatory one-liner. My colleague was depressed for months. Fortunately, he revised and rewrote the paper, and it was subsequently published in a top-tier journal.[2]

In this case, the reviewer was unnecessarily cruel and failed in his responsibility to be constructive. Even worse was the behavior of the journal's editor who forwarded such a review. The editor should have tried a different, potentially more helpful or sympathetic reviewer, possibly advising the original reviewer that this sort of snarky review was definitely not appreciated, nor appropriate.

Now one might ask what a good referee report should look like. There should be three parts. The first part should succinctly summarize the paper demonstrating that the referee understands the paper's principal contributions. The second part should provide suggestions for improvement, if necessary. The third part should be the recommendation for the editor: accept, reject, or the ever popular "revise and resubmit." The referee might also include a suggestion for an alternative journal if the paper does not quite fit the profile of the target journal.

Unconditional Acceptance

Rarely is a paper accepted unconditionally by the first journal to which it is sent. Indeed, it is likely rare that a paper which has been rejected by its first target, but reviewed and rewritten will be accepted unconditionally even at a second target.

Referees are naturally reluctant to accept a paper without adding to or subtracting from it. Unconditional acceptance basically says that in the referee's view the paper is as good as it can be or it might also mean that the referee believes the paper is as good as it needs to be for that particular journal.

In the former case, the decision is made by a confident referee. (This does not mean the reviewer cannot have made a mistake. Confidence is not the same as being correct.) In the latter case, the referee is again confident, but perhaps is lazy. In this case the referee judges that the paper could be better, could hit a better journal, but is good enough for this particular journal. Alas, we will never really be able to tell the difference.

Now, how can an author get an unconditional acceptance? First, the paper can be brilliant. The author might be a wizard. This is rare. Gans and Shepherd (1994) suggest even established wizards do not always get brilliant papers accepted without revision.

Second, the paper may have worked its way through the hierarchy of journals, and the author may have revised the paper along the way on the basis of several helpful reviews.

Third, the author may have presented the paper at an array of top-tier schools and have incorporated the comments of the potential reviewer who might be on the faculty of one such school. In addition, just presenting a paper at a top school and heeding the comments of serious scholars should help a paper along.

Fourth, the author might have aimed at a journal at too low a tier. The author, in search of quick review and speedy publication, might send an "A" level paper to a "B" level journal, perhaps counting on the lazy reviewer noted above.

Unconditional acceptances are rare, but sometimes they ought to happen. Referees who are confident and careful, referees who take a paper for what it is rather than the one they would like to see, should accept appropriate papers without conditions. Referees must resist the temptation to force an author to write the paper the referee would write.

Outright Rejection

This is not as rare as unconditional acceptance, but it is not common. Sometimes, the reviewer will recommend rejection and the editor will follow this advice. More often, a referee who recommends rejection will provide reasons for the rejection with suggestions for what to do with the paper. In this case, the editor can make an informed decision. A kind referee will indicate what might be done to fix a paper and perhaps even suggest alternative journals that might be more appropriate venues for the paper. A kind referee might also advise an author if the paper is a dead-end and why. Not every idea is a good one, and good referees should tell this to authors to keep them from wasting their time on fixes that will lead nowhere.

Of course, referees can be wrong. So depending on the author's stamina, perhaps a "no" should not be accepted as a final decision. As Gans and Shepherd (1994) report, many very good or even great papers suffered rejection at one or more journals. I can speak personally to this.

Although I have no great papers, I do have some that were reasonably good. One, in particular, comes to mind. My coauthor and I experienced four rejections. Two rejected the paper because it was trivial and obvious; two rejected the paper because it was wrong. A paper can be obvious or it can be wrong, but it cannot be both. We persisted and among other citations, the paper was favorably cited in a Nobel Prize acceptance address.

Why should a referee reject a paper? Essentially, either because of faulty analysis, or a poor idea, or it is so poorly executed that the only real remedy is to start from scratch.

Ivo Welch's (2013) blog is also informative regarding referees and outright rejection. He lists four bad reasons for rejecting a paper:

1. "I already knew this." (As a referee if you believe this indicate why. Cite the relevant literature. Do not rely upon what you and your colleagues discuss over coffee at the faculty club.)
2. "The authors were careless." (Referees should distinguish between true carelessness, e.g., sloppy interpretation of important statistics versus minor nitpicks, e.g., misspelling a name.)
3. "This is obvious." (This may or may not be a good reason to reject. But, what might be obvious to the referee after reading the paper may not be obvious to others who have not read the paper. What is "obvious" is water is wet; what is not "obvious" is whether it is hot or cold. As a referee you should explain why something is obvious.)
4. "I don't find it interesting." (Again, maybe this is a good reason to reject. However, this is what the journal editor is supposed to decide based upon the report of a trustworthy and careful referee.)

Lastly, remember as a referee you are charged with being both critic and coach. So even with horrid papers, the referee should try to be helpful even if the advice is to "stop working on this idea."

Revise and Resubmit

This is the most common of referee's suggestions on the first round of review. There are a number of reasons why a referee responds in this way; unfortunately, not all of them are honorable.

Among those that are honorable there are at least three very good reasons for a "revise and resubmit." First, the referee may prod the author to sharpen the argument. For instance (and very superficially) an author might use the word "may" when the real word ought to be "does" or "will." Authors sometimes equivocate, when no equivocation is warranted.

Second, perhaps the referee detects a missing link in the argument. Here the referee might suggest additional statistical work or a more tightly reasoned narrative. Alas, sometimes if the author does what the referee asks, the paper's original thesis may be undone. That is, the referee has discovered a fatal flaw. Good referees should always be aware of this possibility, and so should push the author to explore the missing links.

Third, perhaps the paper's writing quality is lacking. This can happen in at least one of two ways. First, the author may not have placed the paper in the context of extant literature on the subject. A good referee will note this and push the author to do so. The referee might even help a bit by suggesting that the author mention and briefly summarize articles X, Y, and Z. Second, maybe the paper is just poorly written and could be enhanced by a really good literary editor. Here a good referee might suggest new wording or reorganization or even that the author seek the help of an editorial professional. The referee has to be mindful of the fact that the target journal editor's focus is on content, not writing per se.

Now for the less honorable reasons that referees may offer a "revise and resubmit." First, there may be a touch of misplaced kindness. The referee may read the paper, decide it really is not up to the journal's standards, but is too soft-hearted to say so. In this case, the referee may devise a long list of proposed revisions with the hope that the author seeks a different outlet and the paper never shows up in the reviewer's in-box again. Sometimes the author will try to comply; here the redone paper may still not pass muster. But the poor author, misled by a soft-hearted referee, will have wasted precious hours. Good referees should never offer false hope.

Second, a referee may not have time to review a paper as thoroughly and carefully as it deserves. But, wishing to stay in the good grace of an editor, sends an abbreviated review, but asks for some more analysis in order to buy some time. As I noted above, if a referee does not have time for a review, he or she should not accept the request.

Third, a referee, especially inexperienced scholars, may love a paper, but are embarrassed to return it to the editor without asking for additional work. The referee may feel the editor will not give credit for having done a careful review if additional analysis is not demanded. We can only hope that as these referees mature, they will grow in competence and confidence and not demand additional work if it is unnecessary. When this happens the referee creates busy work for the author and extra work for the journal editor who must now process a paper twice.

I am sure there are other good or bad reasons for the "revise and resubmit," but these are enough.

Conclusion[3]

Every referee has to bear in mind the elements of scarcity, the economic issues, or refereeing. There is a time constraint. Try to help the editor keep a modest backlog, so that the editor's successors can leave their own imprints on the journal. Do not try to force a barely satisfactory paper into a top-tier journal because you like the topic and the author treats your own work with exceptional respect. For the most part this sort of management is the editor's problem, but a conscientious referee can help.

Similarly, a careful responsible referee recognizes the other costs of too ready an acceptance. Again, the editor has the final word, but a good referee will provide substantive guidance.

In concluding, note that the "golden rule" for a referee is the same as it is for all of us. Paraphrased, it says, "Treat every paper you are asked to referee as you wish your papers were refereed." It also provides every editor with reviews that you would wish to receive if you were the journal's editor. This pretty much says it all.

Endnotes

1. A very succinct description of the referee's role can also be found in Kelli blog (Cruz, 2011) on *"Guidelines for Refereeing Journal Articles,"* Feb. 8.
2. A very interesting section appears in Friedland et al. (2002) which discusses the deceased Nobel Laureate George J. Stigler as a journal referee. He could be quite acerbic, but generally helpful.
3. For readers wishing to read more about referee behavior, I suggest Hammermesh (1994), Lynch (1998), and Welch (2014) for starters.

References

Cruz, K. (2011). *"Guidelines for Refereeing Journal Articles,"* Kelli blog, February.

Friedland, C., Goodwin, C., Hammond, C.H., Hammond, J.D., Levy, D., Medema, S.G., Naples, M.I., Samuels, W.J., and Stigler, S.M. (2002). Remembrance and appreciation roundtable, George J. Stigler (1911–1931); scholar, father, dissertation advisor, referee, textbook writer, and policy analyst. *American Journal of Economics and Sociology,* 61(3, July): 609.

Gans, J.S. and Shepard, G.B. (1994). How are the mighty fallen: Classic articles by leading economists. *Journal of Economic Perspectives,* 8(1, Winter): 165–179.

Hammermesh, D.S. (1994). Facts and myths about refereeing. *Journal of Economic Perspectives,* 8(1, Winter): 153–163.

Largay, J.A. III. (2001). Three Rs and four Ws. *Accounting Horizons,* 15(1, March): 71–72.

Lynch, J.G. (1998). Presidential Address: Reviewing. *Advances in Consumer Research,* 25: 1–6.

Thomson, P. (2012). *"Refereeing a Journal Article: Part 1: Reading, Part 2: Making a Recommendation, Part 3: Writing the Feedback,"* http://patthomson.wordpress.com/2012/01/07/refereeing-a-journal-article.

Welch, I. (2013). *"CFR Editor's Web Page."* December.

Welch, I. (2014). Referee recommendations. *Review of Financial Studies,* RFS Advanced Access, May 27, pp. 1–32.

13

WRITING SCIENTIFIC JOURNAL ARTICLES

Motivation, Barriers, and Support

JOANNA O. PALISZKIEWICZ

Contents

Introduction

The ability to comprehend and communicate one's ideas in a written text is critical in academic life. Writing is a recursive process involving both cognitive and metacognitive skills (Larkin, 2009). Sound writing is critical for academic and vocational achievement (Graham and Perin, 2007). Writing consists of lower-order skills such as transcription and spelling, and higher-order skills such as ideation. Ideation includes complex, high-level cognitive processes such as planning, translating, and reviewing (Hayes and Flower, 1980; Scardamalia and Bereiter, 1986).

Murray (2002) suggests that most academics have not received any formal training in academic writing; they tend to develop their own skills through a process of trial and error. According to Keen (2007), there is a significant need to consider methods for supporting staff development in regard to writing for publication.

The aim of this chapter is to show motivation, challenges, barriers, and support strategies that are connected with academic writing for authors whose first language is not English. This chapter is organized in the following manner: first extrinsic and intrinsic motivators to publishing are described. Second, the barriers in academic writing are discussed. Next, supporting strategies for authors are outlined. Finally, conclusions and recommendations for future research are made.

Extrinsic and Intrinsic Motivation to Publish

The development of writing for publication is important, because the support of academic writing may increase research productivity (Lee and Boud, 2003), increase faculty esteem (Baldwin and Chandler, 2002), and enhance student writing skills through faculty development. Various forms of motivators can influence people's engagement with academic writing. The motivation to write articles by academics can be divided into extrinsic and intrinsic. For extrinsic motivation, the following factors can be indicated:

- Writing is an important requirement to gain promotion.
- Writing is an important skill for many professions.
- Good writing skills can help people find employment.
- Articles help funding for scholarship and research projects.
- Journal publications can stimulate debate and suggestions for future development and innovation.
- A network of people with similar work can be established.

The majority of academics are required to write research papers for publications in journals because it largely determines their success. The quality and number of the published papers influence promotion and tenure. It is considered essential to the advancement of the profession (Driscoll and Driscoll, 2002; Nelms, 2004) and important to career development (Hollis, 2001; Burnard, 1995; Taylor, Lyon, and Harris, 2004). Furthermore, it provides opportunities for one's employment (Deci and Ryan, 2008). As an example, articles published in high-quality journals are very important for application for scholarship and research projects being funded by the European Union programs. Published articles can stimulate debate and help find new ideas for interesting research. It is also very important in creating networks and sharing knowledge among people, either from the same areas of interest or from other fields of study.

For intrinsic motivation, the following factors can be applied:

* Writing is used to better understand the topic.
* Publications influence the fact that academics gain increased confidence and visibility.
* Writing articles brings about self-motivation to reach one's fullest potential.
* Writing can be enjoyable.

Studies in the area of writing-to-learn provide evidence that writing can contribute to learning (Applebee, 1984). According to Mason (2001), writing serves as a tool to reason, monitor, and communicate conceptions and understandings. Writing provokes new ideas; therefore, it enables and increases the knowledge base in a specific field. Sound writing affects the image of the researcher and in turn, builds the researcher's confidence to develop his or her full potential. Peers can critically review the work of a researcher and provide feedback. The feedback can be used to improve the work of the author. Writing can be an enjoyable task for many.

It is important to focus on promoting levels of extrinsic and intrinsic motivation as they can lead to a higher quality of work being produced, which in turn constitutes positive benefits for academics, students, and the university as a whole.

Barriers to Academic Writing

The literature has documented many barriers to writing (Driscoll and Driscoll, 2002; Nelms, 2004; Nicolaidou, 2012). The most prominent factors that academics can see as barriers to writing articles are

* Language difficulties and cultural challenges
* Lacking confidence in writing ability
* Fluctuating self-belief
* Achieving poor results in publishing: rejected papers
* Lacking knowledge about referencing and avoiding plagiarism
* Meeting deadlines
* Poor engagement of peers/support networks
* Lacking quiet space to work, either at work or at home

To participate successfully in a community, one must learn to communicate in a manner that is approved and accepted by that group (Hyland, 2006). To participate in the international research arena, many scholars for whom English is not their first language feel the pressure to publish in English (e.g., Lillis and Curry, 2006) instead of disseminating their research in their own language. English has been identified as today's academic language (Schuler, 2014). This is also a cultural challenge because academics whose first language is not English need to acquire skills, not only in using the wide range of vocabulary required in their field, but also adopting the range of scholarly conventions used in those disciplines and specific academic writing genres, which causes confusion for both "home" and international writers (e.g., Hamill, 1999).

Lacking confidence in one's writing ability is a barrier. Driscoll and Driscoll (2002) asserted that authors might question whether they have the ability to write for publication. This problem was also noted by Nelms (2004) and Murray (2005).

From the sociocognitive perspective, individuals are understood to possess self-beliefs that enable them to exercise a measure of control over their thoughts, feelings, and actions (Nicolaidou, 2012). Achieving poor results in publishing (rejected papers) can influence one's self-belief.

Authors sometimes lack the skills of citing references in their papers. This is a skill that can be learned and followed easily. In addition, novice authors may not understand the concept of plagiarism. Plagiarism must be taken seriously and in some cases, education is needed to grasp the concept fully and how to avoid it.

Stress and pressure to meet self-imposed (or otherwise) writing deadlines can weigh heavily on the author. Also, lacking a quiet space to work either at home or at the office can be a deterrent in the writing process. In addition, family commitments can rightfully take time away from writing (Timmons and Park, 2008).

The above barriers decrease one's motivation to put energy into writing articles, thus creating an environment that is not productive. The university must make every effort to use strategies that support academics to overcome these barriers.

Supporting Strategies for Authors

In addition to the internal and external motivators, and despite the barriers to effective academic writing, supporting strategies can be identified that can be referred to as enablers of competent academic writing. These strategies comprise various means and mechanisms that facilitate academic writing by providing guidance, support, and encouragement to academics to improve their writing. These factors include:

- Communicating clear guidelines to the academics by the editors of the journals
- Providing social support to academics
- Promoting collaborative writing among academics
- Enabling access to relevant writing support resources and materials
- Providing opportunities for academics to attend workshops, seminars, or courses that would sharpen their writing skills

Most journals require a specific layout and writing style. These requirements are often posted on the journal's website. According to Ellard (2001), many authors do not follow the guidelines provided by journals. This perhaps indicates that the provision of information alone is insufficient to support potential authors. Therefore, other strategies may be needed to develop writing skills, for example, providing social support. Baldwin and Chandler (2002) described four categories of social support in working academics: informational, instrumental, emotional, and appraisal. Access to the materials that include how to write articles can be seen as an informational support. Emotional and instrumental support requires the direct intervention of a third party. This can be in the form of building confidence and self-esteem or the provision of time and resources to write. Appraisal support involves providing feedback to an individual. Collaborative writing strategies help divide the workload and facilitate the development of inexperienced writers through the opportunity to work with more experienced authors (Hollis, 2001; Albarran and Scholes, 2005). There is also a potential for collaboration, helping to overcome the traditionally perceived professional barriers, that is, the use of academic and business partnerships or multidisciplinary partnerships. Numerous advantages of this strategy exist, including the sharing of knowledge, discussing and developing writing practices, mentoring, and obtaining feedback. For others, it is also important to use relevant writing support materials. Nowadays, there is a lot of literature about writing articles for academic journals (e.g., Murray, 2009) and textbooks about writing strategies (Olsen, 2014; Cottrell, 2008). Furthermore, there are various models for writing development, such as one-to-one writing tutorials (Borg and Deane, 2011), group sessions, short courses, and writing groups (Rickard et al., 2009; Jackson, 2009).

The following three steps are often used by authors from conceptualizing to submitting a paper for publication:

1. Preparation
   - Identifying the author or authors, considering the target audience, selecting an appropriate journal, reading the guidelines for authors
2. Writing
   - Planning the structure and outline, selecting appropriate content, adapting and revising based on the guidelines
3. Submitting according to instruction
   - If needed after review, making changes and resubmitting amended manuscript

The amount of time and effort devoted to multiple drafting and revisions is critical to the quality of the finished product (Hayes and Flower, 1986).

Studying the existing literature is important, but gaining practical skills and attending academic writing courses would also be valuable. Academic writing courses are very important in developing writing skills. The courses should focus on three central themes: confidence, writing, and publishing.

Conclusions and Recommendations

The literature reviewed indicates that there is significant and increasing pressure for academics to publish. There seem to be a number of variables that determine an author's academic success with writing that leads to various benefits such as promotion/tenure, self-confidence with writing, and developing critical analysis skills.

The conclusions from this literature review are as follows:

- The development of writing for publication is important because it influences the growth and development of the writer (the academic).
- There are extrinsic motivators, which encourage people to write. Publishing quality articles enables one to be promoted (one hopes), increases the reputation and visibility of the author and associated institution, expands his or her employment opportunities, helps secure funding for research and scholarship projects, stimulates debate and suggestions for future development and innovation, and helps establish a community of scholars that support each other.
- There are intrinsic motivators that enhance academics to write: for example, to gain increased confidence, to improve and reach one's fullest potential, and to enjoy writing.
- Authors may encounter several barriers to writing articles: for example, language difficulties and cultural challenges, lack of confidence in one's writing ability, fluctuating self-belief, a poor track record of quality publications, lack of knowledge about referencing and possible unintentional plagiarism, stress and pressure to meet deadlines, poor engagement of peers/support networks, and lack of a quiet space at work or at home.
- The following supporting strategies were proposed to overcome the barriers: preparing clear assignment guidelines by the editors of the journals, providing opportunities for academic writing, social support, promoting collaborative writing, giving access to relevant writing support resources and materials, and providing opportunities to attend writing workshops/seminars/courses.

Finally, universities should adopt a deliberate plan to support scholars for whom English is not their first language. The plan should include courses, seminars, or workshops that include the fundamental and advanced skills and knowledge authors need to produce quality research papers.

References

Albarran, J.W. and Scholes, J. (2005). How to get published: Seven easy steps. *Nursing in Critical Care*, 10(2): 72–77.

Applebee, A.N. (1984). Writing and reasoning. *Review of Educational Research*, 54: 577–596.

Baldwin, C. and Chandler, G.E. (2002). Improving faculty publication output: The role of a writing coach. *Journal of Professional Nursing*, 18(1): 8–15.

Borg, E. and Deane, M. (2011). "Measuring the Outcomes of Individualised Writing Instruction: A Multilayered Approach to Capturing Changes in Students' Texts." *Teaching in Higher Education*. Retrieved 01.08.2014 http://dx.doi.org/10.1080/13562517.2010.546525

Burnard, P. (1995). Writing for publication: A guide for those who must. *Nurse Education Today*, 15:117–120.

Cottrell, S. (2008). *The Study Skills Handbook*. Basingstoke, UK: Palgrave Macmillan.

Deci, E.L. and Ryan, R.M. (2008). Facilitating optimal motivation and psychological well-being across life's domains. *Canadian Psychology*, 49(1): 14–23.

Driscoll, J. and Driscoll, A. (2002). Writing an article for publication: An open invitation. *Journal of Orthopedic Nursing*, 6: 144–152.

Ellard, J. (2001). How to make an editor's life easier. *Australian Psychiatry*, 9(3): 212–214.

Graham, S. and Perin, D. (2007). *Writing Next: Effective Strategies to Improve Writing of Adolescents in Middle and High School*. Washington, DC: Alliance for Excellent Education.

Hamill, C. (1999). Academic essay writing in the first person: A guide for undergraduates. *Nursing Standard*, 13(44): 38–40.

Hayes, J.R. and Flower, L. (1986). Writing research and the writer. *American Psychologist*, 41: 1106–1113.

Hayes, J.R. and Flower, L.S. (1980). Identifying the organization of writing processes. In L. W. Gregg and E. R. Steinberg (Eds.), *Cognitive Processes in Writing*. Hillsdale, NJ: Erlbaum, pp. 3–30.

Hollis, A. (2001). Co-authorship and the output of academic economists. *Labour Economics*, 8: 503–530.

Hyland, K. (2006). *English for Academic Purposes: An Advanced Resource Book*. London/New York: Routledge.

Jackson, D. (2009). Mentored residential writing retreats: A leadership strategy to develop skills and generate outcomes in writing for publication. *Nurse Education Today*, 29(1): 9–15.

Keen, A. (2007). Writing for publication: Pressures, barriers and support strategies. *Nurse Education Today*, 27: 382–388.

Larkin, S. (2009). Socially mediated metacognition and learning to write. *Thinking Skills and Creativity*, 4(3): 149–159.

Lee, A. and Boud, D. (2003). Writing groups, change and academic identity: Research development as local practice. *Studies in Higher Education*, 28(2): 187–200.

Lillis, T.M. and Curry, M.J. (2006). Professional academic writing by multi-lingual scholars: Interactions with literacy brokers in the production of English medium texts. *Written Communication*, 23(1): 3-35.

Mason, L. (2001). Introducing talk and writing for conceptual change: A classroom study. *Learning and Instruction*, 11, 305–329.

Murray, R. (2009). *Writing for Academic Journals*. Buckingham, UK: Open University Press.

Murray, R. (2005). *Writing for Academic Journals*. New York: Open University Press McGraw Hill Education, Berkshire.

Murray, R. (2002). Writing development for lecturers moving from further to higher education: A case study. *Journal of Further & Higher Education*, 26(3): 229–239.

Nelms, B.C. (2004). Writing for publication: Your obligation to the profession. *Journal of Pediatric Health Care*, 18: 1–2.

Nicolaidou I. (2012). Can process portfolios affect students' writing self-efficacy? *International Journal of Educational Research*, 56: 10–22.

Olsen, L. (2014). "Guide to Academic and Scientific Publication, How to Get Your Writing Published in Scholarly Journals, Academia." Retrieved 01.08.2014 http://www.proof-reading-service.com/guide/index.html

Rickard, C.M., McGrail, M.R., Jones, R., O'Meara, P., Robinson, A., Burley, M., and Ray-Barruel, G. (2009). Supporting academic publication: Evaluation of a writing course combined with writers' support group. *Nurse Education Today*, 29(5): 516–521.

Scardamalia, M. and Bereiter, C. (1986). Research on written composition. In M. C. Wittrock (Ed.), *Handbook on Research on Teaching*. New York: Macmillan, pp. 778–803.

Schuler J. (2014). Writing for publication in linguistics: Exploring niches of multilingual publishing among German linguists. *Journal of English for Academic Purposes*, 16: 1–13.

Taylor, J., Lyon, P., and Harris, J. (2004). Writing for publication a new skill for nurses? *Nurse Education in Practice*, 5: 91–96.

Timmons, S. and Park, J. (2008). A qualitative study of the factors influencing the submission for publication of research undertaken by students. *Nurse Education Today*, 28(6): 744–750.

14

LOST IN TRANSLATION AND OTHER CHALLENGES OF NEW AND INTERNATIONAL RESEARCHERS SEEKING PUBLICATION

ANTHONY K.P. WENSLEY

Contents

Having been an editor of an international journal for over 20 years, I have experience in the challenges associated with developing and publishing scholarly material. In this chapter, I present some of the insights I have gleaned over the years, both with respect to achieving success in publishing and also in managing your scholarly career.

In the early years, although academic publishing was important, there were relatively few journals, submissions were manageable, many universities worldwide saw importance in research, and the "bar" for achieving tenure and promotion at many institutions made reasonable demands with respect to research productivity. Over the intervening years, in many ways starting with the Research Assessment Exercise (RAE) in the United Kingdom, research output has become increasingly important. As a result, a variety of metrics, most important of which are impact factors, have been developed to attempt to assess the importance of published research. In addition, there has been an enormous expansion of publication outlets. Some of the present well-respected journals have been with us since the 1990s (and before), whereas some others have rapidly established themselves as premier journals in their field in a relatively short time.

Why Are You Submitting Your Paper?

Of course, one answer to this question is that you want to have it published but more practically you want to inform others of the research you have completed or, in some cases, the research in which you are engaged. Your publications are a communication between yourself (and your coauthors) and an audience. This means that you should clearly identify your audience in your own mind before you write your paper. What level of familiarity with your subject are you assuming? What will interest them? Some of these questions can be answered, at least in part, through a consideration of the papers that have been published in your target journal. You will want to review papers that have already been published in the journal to get an idea as to the degree to which they use specialized language, make assumptions about the knowledge of the reader, and so on. You might also write to the editor of the journal to ask him or her for a recommendation for papers that have already been published that he or she considers are exemplars of the types of papers that represent the best of the best. In some cases you might be able to short-circuit this process if the journal identifies best papers on a regular basis.

With respect to making sure you are writing for an appropriate audience, it is always a good idea to have a person with some of the characteristics of your target audience read your paper. This may well be a colleague whose judgment you trust. It is worth mentioning that ensuring the clarity of the paper in terms of message, structure, consideration of audience, and so on before you submit a manuscript pays considerable dividends with respect to the speed at which your paper will be reviewed and will likely increase the probability that your paper will be accepted.

On a somewhat more somber note, ensuring the clarity of your writing may make you aware of weaknesses in your paper. This may seem a drawback but if you do not locate any weaknesses in your paper good reviewers will, and it is certainly better if you identify and correct weaknesses rather than leaving it to the reviewers. You may also decide to submit your paper to a different journal or, indeed not submit it in its current form and work to improve it before submission.

A Strategic Approach to Developing Your Career

Preparing one paper for submission is one step in the development of your career so it is worth spending some time in considering how your publication strategy should be developed in the context of developing your career. As a first step, you should review the culture of your institution. How important is research to your institution or department? What type of research is expected (and rewarded)? Does your department/institution have an explicit list of journals (outlets) that it recognizes? If it does not have a set of preferred journals, does it have a well-established set of criteria (impact factors, quality rankings, rankings by particular organizations, etc.) that you can use to assess potential journals for your manuscript submission?

However, it is worth noting that publication in leading research journals will always represent a key element of your academic brand. Even if your institution does not have a well-established research tradition, building a quality research portfolio will strengthen your hand in negotiations with your existing institution as well as make you marketable if you choose to move. Still, developing such a research portfolio is a long-term challenge and needs active planning. Clearly not everything can be planned but thoughtful consideration of what to publish and where to publish it will always pay off.

Choosing What to Publish

It is difficult to provide a clear set of instructions as to what aspect of your research you should publish. Most of us likely consider all of our research work worth publishing but, if we are honest, some of it may be somewhat pedestrian and should best be left aside whereas other research likely may make a significant contribution to the research literature. You have to be as ruthless as you can in your assessment of your research. Again, as with much of the other advice I have given, I would suggest that you cultivate colleagues whose judgment you trust to help you in this assessment. It is a difficult challenge but also an essential one because it is also important from the standpoint of developing your career profile (as discussed below). It is worth noting that such assessments may prove invaluable when you need to provide arguments with respect to the value and contribution of your research when it comes to tenure and promotion time or when you want to further develop your brand!

Developing a clear-headed assessment of your current research also allows you to enunciate future research trajectories and potentially identify other individuals or groups who you can work with to extend your research and, thereby, expand your research portfolio and brand.

Developing a Manuscript for Publication

The decision concerning what research to publish and the outlet that you choose to submit to are interdependent decisions. However, let's start with how we might choose a journal for paper submission.

Choosing Where to Publish

We always want to publish in the most prestigious journals but beware: the higher-quality journals may well have longer turnaround times, have reviewers who are less tolerant of less polished papers, and, inevitably have much higher rejection rates. Many of us have experienced having papers subjected to major revisions on a number of occasions with rejection coming at the end. A very frustrating experience! As with the other decisions you make, it is a question of trade-offs. Ideally, you should develop a portfolio of publications, some that you submit to high-quality first-tier journals and others that you submit to somewhat less prestigious journals. As noted above, you should also be guided by the standards of the institution for which you work. Focus is the key. There is relatively little point to publishing in journals that are not recognized by your institution unless you are doing this to satisfy some other objective. For example, you may want to build up a network of colleagues whom you will be able to work with in future. You may want to "try out" an idea for a paper prior to submitting it to a higher-quality journal. If this is the case, you should be careful because it is not good policy to submit research that contains a significant portion that has been published elsewhere. And this is another point worth making. You should be diligent in self-citation—not to an extreme—but you should provide appropriate citations for your earlier work. Beware, however, of creating the impression that you think that you essentially invented a research theme. Make sure you identify others.

There are many available assessments of the quality of different academic journals. Some of these rankings are well established and broad such as the ranking developed by the *Financial Times* which covers business research in general. Other rankings have been developed in specific areas and are often published in journals in specific areas. It is also often the case that the departments you work in have their own informal rankings.

Before submitting your paper to a journal you should review any statements about the type of research that the journal editor(s) consider to be appropriate for publication. Some of these statements will be quite vague but use your best judgment.

Perhaps the most obvious place to start is to review papers that have already been published in your target journal. Focus on those that address similar subject matter or methodological approach. If there is a discrepancy with respect to subject matter, this may well be an indication that the journal you have selected is inappropriate. In this case you can send your paper to the journal and ask the editor to make a judgment as to whether the paper "fits" with the journal. If you cannot find examples of papers that apply the same methodology, this may present a challenge. It may well be that the journal has an expressed bias (or an unexpressed bias) toward, for example, empirical or interpretative research.

One of the central challenges in gaining success in publishing in major academic journals is that, at least for the foreseeable future, they will require submission in English. Furthermore, given the dire lack of experienced editors and the abundance of time-challenged reviewers, many journals will be tempted to reject manuscripts out of hand if the English used in a submission is of low quality. I discuss this issue in more detail later on in this chapter.

Writing a Paper

More than anything you should remember that a paper is essentially telling a story. In a real sense you have an opportunity to distill your paper into the essential elements. You need to be able to provide short answers to the following questions.

What Is the Unique Contribution of Your Paper? What Knowledge Does Your Paper Add to the Existing Stock of Knowledge?   Identify what it is that you are adding to the store of knowledge. You should be able to encapsulate this in one or two sentences. This will form an essential part of the abstract of your paper. Developing such succinct characterizations of your research is also required in the assessment of PhD theses and, indeed, our own work so the skill you develop in doing this is essential to your future career. It does not necessarily matter if the quanta of knowledge that you add is small and somewhat restricted; what is important to ensure is that you are able to identify your contribution clearly and furthermore, that you are careful not to overstate your contribution. Use of statements such as "makes a major contribution to the field" are generally to be avoided as they will often attract reviewers' opprobrium from the start. Be realistic: don't overstate but also do not understate if you really are convinced that your research is pathbreaking. Although in the latter case it is probably better, as indicated above, that you confirm your assessment with trusted colleagues first! You do not want to provide opportunities for reviewers to form a negative view of your work from the beginning. In my experience, reviewers generally work very hard to ensure that they are unbiased but all of us have a tendency to react strongly against overstatement. Of course, your research may be "groundbreaking" and needs to be recognized as such.

Structuring Your Paper   How you structure your paper very much depends on the nature of your research and the methodology you have chosen as the basis for your research. All papers, however, will require an abstract, an introduction, and a literature review. Let me consider these first. It is probably best to write the abstract once you have written the paper. This is a typical journalistic "trick." The abstract should sum up your paper and provide a clear indication as to its contribution, the methodology, and how it fits into the existing research landscape.

Your introduction can make use of some of the wording of the abstract but should not be merely a repetition of the abstract. Be guided from what you see in existing published papers. The introduction is not a summary of the paper—it provides the context of the paper—why it was important to conduct, the context of the research, and an initial consideration as to where the research fits within the research that has already been completed.

Literature reviews are a challenge. Obviously they cannot in any sense be exhaustive. Be guided by the key publications that you used in developing in the first place. You are trying to give your reader a clear indication as to where your research fits. What are the foundations of your research? Where are the gaps in previous research that your research seeks to fill? It is always difficult to decide how comprehensive to be although the length of your literature review should generally not exceed two to three pages. Remember the essential function of a literature review: it puts your research in context, shows that you have conducted what we might call due diligence with respect to existing research, and establishes the essential credibility of your research. Unless, of course, you are writing a paper intended as an overview of existing research in a particular area and an identification of potential research trends your literature review is not the core of your paper!

The remaining structural components of your paper will depend on the type of research addressed by the paper. Rather than attempt an exhaustive consideration of types of research, I consider three possibilities namely, empirical, interpretative, and case based. This is not an exhaustive classification but it should provide the basis for some useful advice. Let us consider each separately starting with papers that are empirical; that is, they describe research that involves the gathering of data in a variety of different ways, its analysis, and the drawing of conclusions.

Empirical Papers   As we have noted before, it is always a good idea to review the way in which empirical papers are presented in your chosen journal. Typically you will first describe how your data was collected. If you collected the data through a survey you will need to discuss how you selected a sample to survey, the characteristics of the sample population, how you developed the questions, how the survey was administered, and any challenges that you faced in conducting the survey. If your data was obtained from existing databases you will be required to describe the provenance of the data. If you used a survey that you designed and administered yourself you should provide the journal with a copy of all the questions. You will often provide these questions in an appendix to your paper.

Following a discussion of how you obtained the data, you will need to discuss how you analyzed the data. Generally speaking, enough empirical papers have been published for you easily to identify a "template" as to how to write this section. Clearly the initial issues that you need to address relate to the validity of the data from a statistical standpoint. Be guided by recently published research. Typically the standards required of statistical analysis become more challenging as the years pass.

Once you have established the validity and reliability of your data, you can then proceed to analyze that data and draw conclusions. Be careful not to go beyond the analysis that you provide but also make the most of your data. If you are working on testing a statistical model, this will sometimes be presented in a separate section before the analysis section or as an integral part of the analysis section.

Interpretative Papers   These types of papers present a challenge because there is always a wealth of data and it is difficult to decide how much of the data to include. Again, as with the empirical papers, you need to describe where the data came from. Often you will have conducted interviews, reviewed documents, and used other extensive data sources. You will then have to discuss how you analyzed the data: how you identified major themes, concepts, and so on. When discussing these themes or concepts, choose the best sources and clearly identify them. Link them to your underlying data sources.

When you are deciding which journal for submission, you should check to make sure that the journal has actually published interpretative research in the past or if it clearly states that this type of research is acceptable.

Case-Based Papers A variety of journals publish both research-based papers and case-based research papers. Generally, research-based papers are considered to be somewhat superior to case-based papers. The structure of case-based papers varies somewhat among different journals so be guided, as I have noted above, by previously published papers.

Typically the penultimate section of your paper is a discussion section. It is here that you present an extended explanation of your analysis. This is particularly important when the analysis is statistical. Do not assume that your readers are necessarily completely familiar with statistical concepts. When you are testing hypotheses clearly explain what you have established. Generally interpretative and case-based papers provide much of their discussion within the body of the analysis section of your paper.

The final section of your paper, the conclusion, should provide a summary of the paper and highlight key findings. For some journals, conclusions are required to address practical as well as research-oriented implications, whereas other journals require a focus on purely research-oriented conclusions. Be guided by the journal and previously published papers.

Language Problem Over the years, I have received many papers that were in need of extensive revisions purely to make them intelligible in English. Many, but not all, of these papers were written by authors who were not native English speakers. There are various strategies to adopt to assist you in improving the quality and style of your papers. One approach is to find coauthors (and coresearchers) who are native English speakers. Of course this means that your research will result in joint publication but better that you achieve successful publication rather than receive rejections. An alternative path would be to take advantage of professional editors. I would suggest that the former approach has more potential as it will allow you to extend and develop your research and build your research network.

It is also worth noting that some journals are more accepting of manuscripts that are grammatically and stylistically challenging as is my own journal. However, as I have noted above, the increase in submissions to journals and the continued predominance of English language journals is likely to make it increasingly important to ensure the quality of the English utilized in your paper.

Assuming that you have passed the first round of reviewing you will receive reviewer feedback. Obviously you should take all the reviewers' comments seriously. In the following sections I take the opportunity to provide some advice.

Responding to Referees   Good referees will generally make it clear how you are to address their concerns. Provide a detailed response as to how you have responded to the issues that they have raised. If you do not consider their concerns to be appropriate provide a clear indication. If reviewers provide contradictory suggestions then choose the one(s) that are consistent with your thinking and, in your opinion, most defensible. If you genuinely have concerns with respect to responding to reviewers you can certainly communicate with the editor although you should do this very sparingly inasmuch as the editor will generally defer to the reviewers unless they are in significant disagreement.

Sometimes editors will indicate that papers need major revisions and express significant concern as to whether such revisions are achievable. In these situations you will have to decide whether it is worth the effort even though the possibility of rejection may be high. Your decision will be influenced by the quality of the journal and the extensiveness of the required revisions. You may again wish to consult with your more experienced colleagues.

How Valuable Is Work with an International Focus?   Although replication is fundamentally important to science, simply replicating a study that has been conducted in one country in another is of relatively little interest to many journals. This may seem harsh but for high-quality journals they will typically be looking for the research to be extended in some way or for the research to demonstrate important differences. For example, either the results differ significantly from previous research or the results are the same as with previous research although the context would lead one to predict that they would not be the same.

Conclusions

It is hoped that these insights from my 20 years of journal editing will be useful to those submitting papers to academic journals. Mentoring junior faculty in terms of publishing, research, teaching, and service should be carried out at colleges and universities. I wish you much success on your publishing journey.

15

PUBLISHING IN LEADING JOURNALS

An Overview for Aspirant Authors Early in Their Careers*

SUPRATEEK SARKER

Contents

---

* This short opinion piece is based on the author's experience as an author, reviewer, and editor in the information systems discipline. His ideas are greatly influenced by leading scholars such as Professors Allen Lee, Carol Saunders, Detmar Straub, and Dorothy Leidner. Needless to say, he takes full responsibility for any of the ideas expressed.

Introduction

Publishing in highly reputed journals is now an imperative for many scholars in order to progress in their academic careers. Many PhD programs are requiring their students to publish in high-quality journals as part of the program. In the recruitment process for starting academic positions (assistant professor, lecturer, etc.), an increasing number of research-oriented universities around the globe are favorably considering candidates only if they have publications in top-tier journals, or, at the very least, if the candidates can demonstrate their ability to publish work in such journals (as evident from revise–resubmit decisions received by them). For universities having rigorous tenure and promotion processes, publications in leading journals (considerations of both quality and quantity) serve as a key input. Even grant agencies in many countries are known to judge the stature and credibility of a scholar applying for funding based on the scholar's list of publications in leading journals. Unfortunately, publishing in these journals in a sustained manner is not among the easiest tasks faced by an academic. Leading journals tend to have high standards, with reviewers and editors often demanding unreasonable levels of theoretical and methodological sophistication, scale of the study, and contributions. Moreover, the review process can be quite tedious and requires a high degree of resilience, with authors having to endure several review cycles before they are able to receive final acceptance (if all goes well), or a rejection decision.

Within such a context, many authors often wonder if publishing in the leading outlets of the discipline is really worth the effort. This is particularly true for authors in educational systems (say, teaching-oriented universities, and university systems in many countries) where there are enormous teaching load and service responsibilities, and the formal requirements for advancement do not specify the list of target journals, but instead ask for "international" journals, or "peer-reviewed" journals, or suitably "indexed" journals. Many researchers also try to achieve the minimum requirement of publishing work in leading journals (per tenure requirements, or requirement for promotion to full professor rank in their system), and then retreat to a publications strategy of contributing to outlets that are not so demanding.

Although I believe that each scholar has good reasons for pursuing scholarship in a given way, my goal in this brief chapter is to touch upon a number of issues related to publishing, so that scholars can be more informed when deciding on their path. I pose five questions that I have wondered about early in my career, and attempt to answer them. Two points are worth mentioning: first, the questions that I address do not necessarily form a coherent set, but I am hopeful that they help in providing a holistic picture about publishing in leading journals; second, many of the answers represent my opinion, shaped by a certain disciplinary context and a certain set of experiences, and not every scholar or editor would necessarily agree with my attempted answers.

Five Questions about Publishing in Leading Journals

Q1: Why Publish in Leading Journals?

As mentioned above, pursuing publications in leading journals is no doubt challenging, and often frustrating; however, it can greatly help in career advancement, especially in research-intensive environments. It can serve as "hard currency" in the academic world, and can enhance scholarly reputation or prestige in the discipline (within and across the individual's home institution), and consequently job mobility globally. Beyond the tangible career rewards, such publications provide wide exposure to the authors' ideas. Such visibility is critical for the members of the discipline to learn about, appreciate, critique, and build upon the authors' work. Lower-tier journals may be attractive to many authors because of the lower degree of resistance of the review process; unfortunately, many of the ideas published in such journals are lost and the contributions tend to go unacknowledged and unappreciated by the wider and more influential audience.

Q2: What Do Highly Leading Journals Look For?

Leading journals aspire to publish work that is *outstanding* in some sense, at least for the discipline. What makes a piece of work "outstanding"? Although there may be different interpretations of the term, from a practical standpoint, it could imply:

- The work is expected to be "*frame-breaking*" or *pathbreaking*, providing a new and potentially useful way to view and examine a phenomenon, and enable researchers in the community to break away from "path-dependent theorizing" and empirical investigations, and pursue new interesting paths (Bansal, 2012). In other words, such work can start a new stream or lay building blocks for a theoretical or methodological tradition, potentially making it a highly citable article (Straub, 2009c), which is of much interest to editors.
- The work is expected to be an outstanding instance of normal science (Kuhn, 1962), where the authors offer contributions to the existing body of knowledge, staying within the existing frame and not straying too far from the path already established. Some of these contributions can be quite significant when judged from within the tradition of research; however, from outside the tradition, the work may not be seen to be all that impressive, and may be characterized as incremental or nuanced.

Given that pathbreaking or frame-breaking work requires creative thinking and sophisticated argumentation often against the flow of mainstream thinking, it is likely that such manuscripts encounter severe resistance in the review process; thus, it is not surprising that such work does not appear in these journals as often as discernible readers might desire. Having a deep knowledge of the trends in the discipline and a high level of credibility can help scholars publish such work.

The kind of work that we see more in leading journals involves work that has excellent data, sophisticated analysis, recognizable theory possibly with a twist, but with contributions that are not particularly novel or exciting (Straub, 2008). Such papers represent puzzle-solving (Kuhn, 1962) with minimal apparent flaws. Doctoral programs frequently gear students for this kind of work through their emphasis on a narrow literature, specific theoretical approaches, and methodological rigor.

Q3: What Are Some of the Characteristics of Papers Published in Such Journals?

The discussion above leads us to the characteristics usually privileged by leading journals, some of which we discuss below.

Nature of the Topic   Obviously, any topic on which a researcher (or a team of researchers) invests significant time is one he or she or the team considers as being important. However, the goodness of the topic is assessed with respect to a specific research community's and, indeed, the specific journal's goals and foci.

There is also a tension that many authors experience when choosing a topic: should they choose a new topic or should they try to seek to conduct and publish work in a mature area? Obviously, currency of a topic (as evident from the world of practice, or popular magazines and trade journals) is valuable, but, if the topic is too new and different from what the community of researchers has been dealing with, the authors are likely to encounter a great deal of resistance in the review process. However, resistance is not necessarily bad. The probability of being successful is negatively related to the level of resistance; however, in the event the paper does get published, it has the possibility of being one of the best-known papers in the area, with a large number of citations, as the body of research on this topic grows. I view these studies as having high risk but potentially high return.

Studies in areas where the work is established but the literature isn't quite saturated probably present the best chances of being published, everything else being held constant. However, these studies also tend to get lost; that is, unless there is something special about the study in terms of the data, theory, or a methodological technique, they suffer from low memorability. Not many researchers in the community pay attention to such papers or use them in their own work.

Studies in areas reaching saturation are relatively more difficult to publish, because the research community believes that not much new or interesting can come out of more research. However, the high-value opportunity under these circumstances is for developing review papers (and sometimes meta-analysis papers) that help readers make sense of a large body of knowledge in the area, and see insights and trends that are not easily evident to those not immersed in the area's literature. Review papers usually require time to mature, before and during the review process, and can thus take significant time to be published. However, once published, these papers become the authoritative, highly cited source for future researchers.

Whatever the nature of the topic one chooses to publish, there are risks and rewards. Although different topics allow for different levels of "wow effect" or enlightenment, authors should keep in mind that submitting a manuscript with some relevant and novel content would generally help in getting a favorable reaction from the review teams in leading journals, all else remaining the same.

Theoretical and Methodological Rigor/Sophistication/Innovativeness   To be published in a leading journal, manuscripts typically must demonstrate a high level of theoretical engagement and methodological rigor (Sarker, Xiao, and Beaulieu, 2013). These two elements are critical in research because data and results without the theoretical foundations are devoid of meaning, and on the other hand, analysis that is methodologically suspect cannot be trusted to validate, inform, or transform theory. Criteria for methodological rigor must be applied according to the nature and goals of the study. Similarly, although there is often no perfect or best theory for any study (Walsham, 1997), the authors should carefully consider candidate theories and, if appropriate, explain why a particular theory may be interesting to study a given phenomenon. Apart from technical correctness, the authors need to know which of the theories and methods are considered to be interesting, valid, or desirable within the research community related to the journal at the time the paper is submitted.

A final observation I would like to make in this regard is that, for most editors, methodological and data-related flaws are of greater concern than theoretical flaws, perhaps because the formulas are difficult to fix through revisions and are potentially a source of greater embarrassment for a journal.

Scale of the Study and the Nature of Data   Many journals do not explicitly acknowledge this issue, but my own view is that it helps to have a study that is the result of a large data collection effort. If nothing else, it signals ambition and hard work. Of course, the scale of studies cannot compensate for fatal flaws in other areas. In addition to the scale, the nature of the data—the level of appropriateness to the study and the relative difficulty in obtaining such data—also has a bearing on publication decisions in leading journals.

Presentation   Finally, papers that are written according to the guidelines of the journal and communicate the various aspects of the paper, including the motivation, theory, methodology, discussion, and contributions in a manner that is easily understandable to readers are likely to do well in the review process. This could mean that, for a non-English native speaker hoping to get published in a leading English-language journal, it may be worthwhile to seek the services of a language editor. However, I would like to emphasize that English language skills are only part of the issue. Far more important is the knowledge of expectations on how to present the various aspects of the study, which includes how papers are to be motivated, how data and methodology are described, how analysis is reported, and how contributions are claimed.

For a more detailed coverage on why leading journals accept papers, see Straub (2009b).

Q4: What Is the Nature of the Editorial Process of Leading Journals?

All leading journals have low acceptance rates, indicating that most submissions to the journal are rejected at some point. However, the editorial attitudes toward the low acceptance rates are not uniform (e.g., Straub, 2009a). Some editors see this with a sense of pride, others see this as a "necessary evil" (Straub, 2009a), and yet others as a systemic problem that needs to be addressed—after all, the goal of journals is to publish rather than to reject research papers!

Although the acceptance rates are always going to be low in a leading journal, the issue of interest to a potential author is whether the journal's review process is one of selection, development, or elimination. By *selection*, I mean that truly the meritorious papers are chosen for publication. The role of the selection-oriented review process is to sift through the submissions and identify the special pieces of work. Obviously, the quality and openness of the editors has significant influence on the process of identifying which papers (or type of papers) are meritorious. In contrast, many journals see their role as development of scholarship wherein manuscripts that are seen to have promise are developed through the review process, characterized by *diamond-cutting* (e.g., Saunders, 2005a,b). The idea is that even though the manuscript may eventually be rejected because it fails to meet the expectations of the review team, the authors and the discipline would have gained a great deal from the review process. This is because a much better manuscript would have been created through the collective contributions of the authors and the review team over a prolonged time period of engagement with the research. The manuscript (even if rejected at a given journal) can be submitted to an alternate journal for publication consideration.

Naturally, significant commitment of reviewers, editors, and authors is required for such a process to succeed. This is because the review team has to engage deeply with the work, and not only point out what is not right about the manuscript but offer detailed suggestions on how the problems may be fixed; sometimes, the contribution of the editor and reviewers to a manuscript is not too different from that of a coauthor (see Lee, 1999). Naturally, this process would involve openness and intellectual resilience of the authors, wherein they are able to embrace the suggestions and implement them in the manuscript. Unless managed carefully, the developmental review process can become dysfunctional, in that the review process may go on for several rounds without any guarantee that the paper will eventually be accepted. This might negatively affect the authors who must publish a certain number of articles within a given time frame. The other issue is that, on many occasions, the authors feel that the resulting manuscript no longer represents what they had wanted to share with the community, but that they had ended up producing the paper that was envisioned by the reviewers and the editors.

Finally, some review processes are geared toward elimination. The goal of the editorial process is to minimize the chances of an undeserving paper being published in the journal (Straub, 2008). Within such a context, usually manuscripts with incremental contributions that are (methodologically) defensible tend to fare better. Papers that introduce new and interesting material for which there are no clear standards of evaluation or strategies for defense are usually unable to survive the review process. This is particularly true when the editor(s) are not closely involved in the review process and largely rely on reviewer recommendations to make publication decisions.

Q5: How Do Researchers Approach the Publication Process?

This is not an easy question to answer, because every researcher has her own unique approach or strategy that guides her. We have already touched on some of the basic contributing factors such as those of

Table 15.1 Different Modes of Pursuing Publications in Leading Journals—A Simplified View

	INCREMENTAL CONTRIBUTION	SUBSTANTIAL CONTRIBUTION
Early career	"Satisfying constraints"	"Showcasing" unique skills and scholarly potential
Mature career	Winning a "sport"	Seeking "self-actualization"

being patient, determined, and resilient, knowing the research community or audience, developing technically correct/defensible and relatively novel work of impressive scale, and understanding the nature of the editorial process (e.g., Lee, 2000a). In addition, it may be worth looking at the attitudes of authors toward publishing that keep them motivated at different stages of their career. A simplified (or perhaps simplistic) view is provided in Table 15.1.

Essentially, as Table 15.1 shows, an academic career may be divided into two phases: earlier career and mature career. Similarly, the scholarly contributions intended or offered by authors can be classified as being incremental or substantial. A large proportion of authors seeking to publish in leading journals in the early phases of their academic careers tend to view their efforts as "satisfying constraints." In other words, the focus is on learning about the formula for preparing manuscripts that can be successfully navigated through the publication process, and succeeding, at least a few times, in publishing the manuscript, thereby surviving this stage of their career. In fact, those who succeed in going beyond meeting constraints by publishing more often in leading journals than is typical, often come to be acknowledged as leading upcoming researchers in the discipline. In contrast to the attitude of "meeting constraints," a small proportion of early career researchers approach publishing as "showcasing" their unique talents, novel ideas, and research skills. Such an endeavor requires independent, and often a counter-mainstream thinking; consequently, it is risky, but can fast track one's career if one is able to publish such material in leading mainstream journals. It appears that a selection-oriented or developmental editorial process rather than the elimination-oriented editorial process is more suited for authors approaching publications in a "showcasing" mode. In any case, success in showcasing or in meeting constraints modes are both creditable, and can lead to high visibility and favorable career trajectories.

As scholars' careers mature, they tend to pursue some combination of the two modes discussed below in order to continue publishing at the highest level: approaching the publication process as a "sport" or as an opportunity to achieve "self-actualization."*

Those approaching publication as sport tend to find meaning in winning and losing in the publication game rather than in contributing profound or interesting ideas. Often, the focus is not on what is published, but on how many were published, and in which journals. Within this world of publication as sport, productivity in specific journals is the key metric, and many successful scholars spend a great deal of energy in perfecting their skills for the game and in setting up a sustainable publication machinery, with colleagues, doctoral students, sources of data, and so on.

The self-actualization mode, on the other hand, leads authors to formulate manuscripts that can potentially enable transformation in the discipline, about some aspect of research (e.g., theory, methodology, genre of research) or about the scholarly process itself. Success in this form of publishing might lead the author to be viewed as a thought leader within a tradition or for the discipline as a whole. Within the world of the self-actualizing researcher, the key metric, if any, is long-term impact, which is not only recognized with citations but also being acknowledged for starting and solidifying a new tradition/trend in research.

Concluding Remarks

Based on my conversations with many researchers around the globe, I believe that many scholars feel intimidated by the "top journals" and start believing that they will never succeed in publishing in such journals. Thus, they do not even consider sending their work to reputed outlets, unless their system demands it. This is very unfortunate, because many of them have undertaken excellent research that could potentially be of interest to the journals. My advice would be for authors not to eliminate these highly visible journals from the set of journals they submit to, but to follow a "portfolio approach" of

---

* Of course, it goes without saying that many accomplished scholars pursue publications in both the sport and the self-actualization modes by having a portfolio of manuscripts featuring different kinds of research.

targeting work to different level journals, so as to deliberately manage their risks and returns.

Noting the social nature of the review process, it is important for aspiring authors to be aware of the priorities and preferences of the audience, including the editors and reviewers (e.g., Lee, 1999, 2000a,b). Where possible, they should take the opportunity to nominate reputed scholars to serve as reviewers and editors for their work. The feedback they receive may appear harsh initially, but it also offers an enormous opportunity to learn about how the target audience thinks about the phenomenon and about research in general. This learning is not wasted if the particular manuscript is rejected, but can inform the development of subsequent manuscripts. Two other ways of getting to know about the journal are to read several recent issues and to review for the journal (a good start is to offer to review for the journal).

With respect to keeping themselves motivated, authors may find it useful to see if they can identify with the modes depicted in Table 15.1, or hybrid forms of those modes. Each mode has a different set of values and priorities associated with it that may better fit a given researcher's institutional as well as individual context.

In conclusion, I would reiterate the point that publishing in leading journals is rewarding but quite challenging. This is especially so for young scholars who have PhDs from programs that may not be tightly coupled with the research community the focal journals represent. I hope the discussion above not only inspires but also offers pointers for readers who are keen to publish in such outlets.

References

Bansal, P. (2012). Inducing frame-breaking insights through qualitative research. *Corporate Governance: An International Review*, 21(2): 127–130.

Kuhn, T.S. (1962). *The Structure of Scientific Revolutions*. Chicago: University of Chicago Press.

Lee, A.S. (2000a). Editor's comments: Submitting a manuscript for publication: Some advice and an insider's view. *MIS Quarterly*, 24(2).

Lee, A.S. (2000b). Editor's comments: The social and political context of doing relevant research. *MIS Quarterly*, 24(3).

Lee, A.S. (2000c). Editor's comments: Irreducibly sociological dimensions in research and publishing. *MIS Quarterly*, 24(4).

Lee, A.S. (1999). Editor's comments: The role of information technology in reviewing and publishing manuscripts at *MIS Quarterly*. *MIS Quarterly*, 23(4).

Sarker, S., Xiao, X., and Beaulieu, T. (2013). Guest editorial: Qualitative studies in information systems: A critical review and some guiding principles. *MIS Quarterly*, 37(4): iii–xviii.

Saunders, C. (2005a). Editor's comments: Looking for diamond cutters. *MIS Quarterly*, 29(1).

Saunders, C. (2005b). Editor's comments: From the trenches: Thoughts on developmental reviewing. *MIS Quarterly*, 29(2).

Straub, D.W. (2009a). Editor's comments: Diamond mining or coal mining? Which reviewing industry are we in? *MIS Quarterly*, 33(2): iii–vii.

Straub, D.W. (2009b). Editor's comments: Why top journals accept your paper. *MIS Quarterly*, 33(3): iii–x.

Straub, D.W. (2009c). Creating blue oceans of thought via highly citable articles. *MIS Quarterly*, 33(4).

Straub, D.W. (2008). Type II reviewing errors and the search for exciting papers. *MIS Quarterly*, 32(2).

Walsham, G. (1997). Actor-network theory and IS research: Current status and future prospects. In A.S. Lee, J. Liebenau, and J.I. DeGross (Eds.), *Information Systems and Qualitative Research*. London: Chapman and Hall, pp. 466–480.

Index